Fritz Herbert

Atmospheric Trace Constituents

Fritz Herbert (Ed.)

Atmospheric Trace Constituents

Proceedings of the 5th Two-Annual Colloquium of the Sonderforschungsbereich 73 of the Universities Frankfurt and Mainz and the Max-Planck-Institut Mainz, Held in Mainz, Germany, on 1 July 1981

With 61 Figures

V

Friedr. Vieweg & Sohn Braunschweig / Wiesbaden

CIP-Kurztitelaufnahme der Deutschen Bibliothek

Atmospheric trace constituents: held in Mainz, Germany, on 1 July 1981/Fritz Herbert (ed.). – Braunschweig; Wiesbaden: Vieweg, 1982.
(Proceedings of the ... two-annual colloquium of the Sonderforschungsbereich 73 [Seventy-three] of the Universities Frankfurt and Mainz and the Max-Planck-Institut Mainz; 5)

ISBN-13: 978-3-528-08523-0 e-ISBN-13: 978-3-322-90097-5
DOI: 10.1007/978-3-322-90097-5

NE: Herbert, Fritz [Hrsg.]; Sonderforschungsbereich Atmosphärische Spurenstoffe ⟨Frankfurt, Main; Mainz⟩: Proceedings of the ...

1982

Softcover reprint of the hardcover 1st edition 1982

Produced by Lengericher Handelsdruckerei, Lengerich

Foreword

This colloquium on *Atmospheric Trace Constituents* held in Mainz on July 1, 1981, was a joint one-day meeting of all scientific working groups of the SFB-73. The purpose of this meeting was to provide a fairly intensive and comprehensive review of the SFB-73 research work during the last two years and to discuss the state and the problems as well as possible future developments of the various projects.

According to the scientific program of the SFB, the meeting consisted of two separate sessions devoted to actual research problems of trace gases (session I) as well as aerosol particles (session II) in the atmospheric environment. As in previous years, each of the 14 partial projects presented its efforts at least by one research contribution. In nearly twenty papers and discussions, methods, results and the present state of art as well as particularly numerous open questions in the experimental and theoretical studies could be explained. The range of topics considered can be gauged from the contents of this publication. The proceedings contain almost all papers presented at the colloquium.

These papers have been issued as quickly as possible. Most of the authors devoted considerable efforts to keep within tight time and space deadlines. It is a pleasure to acknowledge these efforts here. A few number of papers, however, could only be received with a considerable time extension and this unfortunately caused an editorial delay of several months.

As always, in an endeavour such as this, the efforts of a number of other people are necessary for its accomplishment. In particular, thanks are due to all colloquium participants (first of all to the SFB-speaker H. W. Georgii who organized the session as well as C. Junge and P. Warneck who agreed to act as chairmen) for making the meeting itself to a success. In this regard also the hospitality given by the Max-Planck-Institute Mainz and the technical assistance of some co-workers at the Institute of Meteorology and Geophysics of the University Frankfurt, is gratefully acknowledged. Especially, thanks are due to Mrs. R. Kühn (secretary) for her expert help in preparing the manuscript.

Frankfurt, April 1982

F. Herbert
Institute of Meteorology and Geophysics
University Frankfurt, F. R. G.

Sonderforschungsbereich 73
'Atmospheric Trace Constituents'

Members:

S. Beilke, Univ. Frankfurt
F. J. Comes, Univ. Frankfurt
P. Crutzen, MPI Mainz
R. Eiden, Univ. Mainz
H. W. Georgii, Univ. Frankfurt
K. O. Groeneveld, Univ. Frankfurt
G. Hänel, Univ. Frankfurt
H. J. Hahn, MPI Mainz
F. Herbert, Univ. Frankfurt
R. Jaenicke, Univ. Mainz
W. Jaeschke, Univ. Frankfurt
M. Schidlowski, MPI Mainz
W. Seiler, MPI Mainz
P. Warneck, MPI Mainz
W. Zdunkowski, Univ. Mainz

Projects:

Physico-chemical Processes of the atmosphere (E)
Trace Gases (F), Aerosols (G), Paläoatmosphere (H)

Organization of the Colloquium:

H. W. Georgii

Contents

Session I:
Trace Gases

Measurements of Reduced Sulfur Gases in the Atmosphere

H. Bingemer

Institut für Meteorologie und Geophysik der Johann Wolfgang Goethe-Universität, Frankfurt/Main, FRG

A b s t r a c t

The role of biogenic reduced sulfur gases in the tropospheric cycle of sulfur has so far been largely based on mass balance considerations, but only few measurements of the gases.
We have established a method for preconcentration and subsequent gas chromatographic and flame photometric analysis of reduced sulfur gases from the atmosphere. First measurements in environments of various degrees of pollution are reported. On a cruise across the Atlantic, samples from the marine atmosphere habe been collected and analyzed on board ship immediately after collection.The concentrations of dimethylsulfide in the atmosphere ranged from 2 to 44 ng S(DMS)/m^3. Model calculations for the transfer of DMS across the air-sea interface suggest a global flux of 20-33 Tg S(DMS) per year from the oceans to the atmosphere.

1. INTRODUCTION

Most mass-balance models of the atmospheric sulfur cycle require a flux of biogenic reduced sulfur gases into the atmosphere, mainly from marine sources, to balance known sinks. The estimated source strenghts of the oceans and continents are based predominantly on circumstantional reasoning, i.e. mass-balance considerations, but on few data on reduced sulfur gases in the oceans and atmosphere.
The fluxes required to close the balance vary beween 170 TgS per year (marine sources only) in earlier estimates (Eriksson, 1963) and 2-34 TgS per year in more recent estimates (Janßen-Schmidt 1981, Granat 1976).
It is known that living biota as well as decaying organic matter release methylated sulfur (Bremner and Steele,1978).

Dimethylsulfide (DMS) is released by both benthic and planctonic marine algae (Challenger,1951, Andreae et al.,1981). It is therefore present everywhere in the surface ocean (Lovelock et al.,1972) and, due to its volatility, presents a potential carrier of sulfur to the atmosphere.
DMS can be oxidized in the atmosphere by a number of mechanisms. The reactions considered to be the most important are those initiated by hydroxyl radicals (Atkinson et al.,1978; Kurylo,1978). SO_2 is believed to be a major product of this reactions (Cox and Sheppard, 1980; Graedel,1977).
We have measured the concentrations of dimethylsulfide, methylmercaptan and hydrogen sulfide in polluted and remote continental atmospheres. On a cruise of the research vessel METEOR across the Atlantic samples from the remote marine atmosphere were taken. On this cruise the concentration of DMS dissolved in ocean surface water has been determined by M.O. Andreae and W.R. Barnard from Florida State University (see Barnard et al.,1981).
Representative information on the concentrations in different environments will be presented. The diurnal variation of DMS will be dicussed to aid in the understanding of its atmospheric chemistry. From the concentration gradient of DMS across the air-sea-interface the flux of DMS from the ocean into the boundary layer of the marine atmosphere is estimated.

2. METHOD

The concentration of reduced sulfur gases in air is determined by preconcentration and subsequent GC/FPD analysis. For preconcentration the air is drawn at a rate of 0.5-2 l/min through an 8 mm I.D. quartz tube filled with gold wool made of 50 µ gold wire or 15 µ gold foil.
Reduced sulfur gases are concentrated on the gold surface by chemisorption (Braman et al.,1978; Ammons,1980). Sulfur dioxide in the air is removed by a scrubber tube (10% Na_2CO_3 on Chromosorb W-AW-DCMS) in front of the sampling tube. The volume of the air sample is measured by a mass-flowmeter and integrator.
Immediately after collection of the sample, the gold tube is attached to the GC/FPD system (Fig. 1).

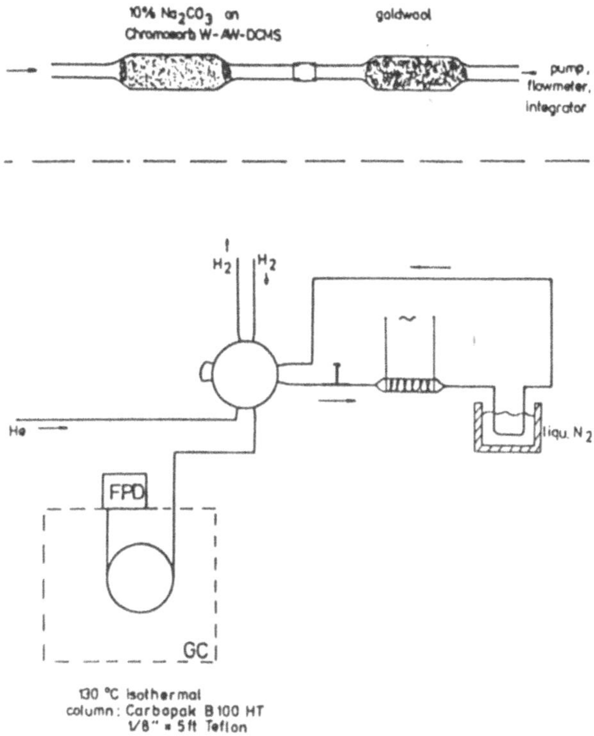

Fig. 1 System for preconcentration and analysis of reduced sulfur gases

By means of a six-way sample injection valve the helium carrier flow is diverted through the sampling tube. After injection of o.1 ml ethyl iodide vapor the sampling tube is heated for desorption. Hydrogen sulfide and methyl mercaptan are converted to their ethyl derivatives, while DMS remains unchanged. The compounds released in the helium carrier flow during the five minute desorption period are trapped again in a capillary trap immersed in liquid nitrogen. Upon removal of the liquid nitrogen and warmup of the trap they are transferred to the Carbopak B 100 column for separation and subsequent detection.
The method is calibrated by taking samples from a calibration gas stream, going through the whole procedure of sampling and analysis as described above.

Calibration mixtures are generated with commercial permeation tubes (Dynacal, KINTEK), using a high precision temperature control system, and dilution with compressed, clean air.

3. RESULTS

The concentration of the reduced sulfur-gases in the atmosphere (Table 1) generally is in the lower ppt-range, except for the samples taken in the vicinity of a chemical plant processing pulp and viscose. In the city of Frankfurt H_2S represents the major reduced sulfur pollutant with average concentrations of several hundred nanograms per cubic meter. The mean concentrations of DMS and CH_3SH are 32 ng S(DMS)/m^3 and 9.3 ng S(CH_3SH)/m^3 respectively. Concentrations are highly variable at short time scales (Fig. 2), due to advection of polluted air.

Table 1

sampling site environment	concentrations ng S (of compound)/m^3 STP
Wiesbaden 1.5 km from a chemical plant processing pulp and viscose	H_2S: 100-5000, mean: 1918 CH_3SH: 20 - 570, mean: 180 DMS: 35 - 635, mean: 198 (10 samples)
City of Frankfurt polluted, 15 km from major industrial sources	DMS: 5 - 470, mean: 32 (93 samples)
Mt. Kleiner Feldberg 820 m MSL, moderately polluted	DMS: 5 - 21, mean: 9 (25 samples)
North- and South Atlantic clean marine air	DMS: 1 - 44, mean: 4 (109 samples)

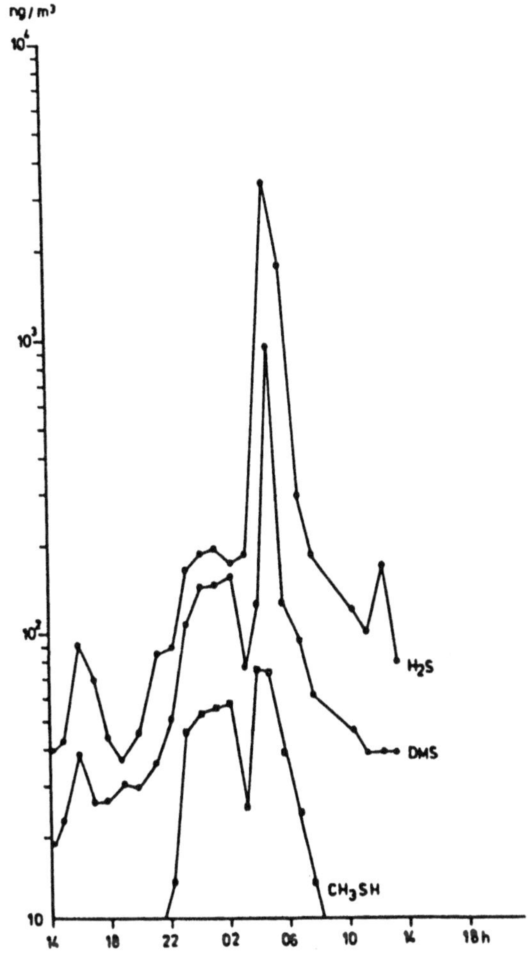

Fig. 2 Concentration of reduced sulfur gases in the city of Frankfurt on 13/14 June 1981

For atmospheric DMS, a characteristic feature is the relative small variation of the average concentration among environments of different degrees of pollution. This primarily displays the effect of the short atmospheric lifetime of DMS and points out the dominance of continuous widespread natural sources, as compared to anthropogenic emissions.

3.1 DMS IN THE MARINE ATMOSPHERE

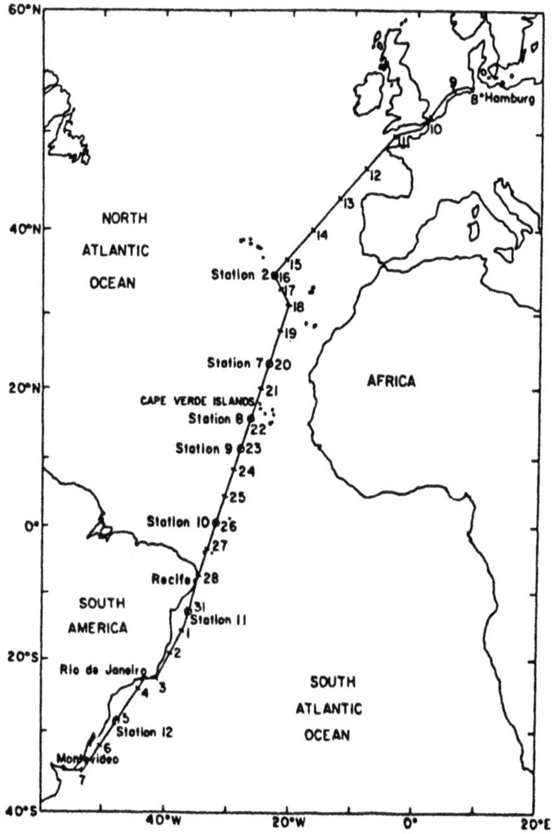

Fig. 3 Cruise track of RV METEOR from October 8, to November 7, 1980

The concentrations of DMS observed in marine air during the Meteor 56/I cruise (Fig. 3) fell in the range from less than 1 ng S(DMS)/m^3 (the detection limit of the method) to 44 ng S(DMS)/m^3. The average concentration for the whole cruise was 6.1 ng S(DMS)/m^3 (72 samples).
Figure 4 shows the concentration of atmospheric DMS along the cruise track from 54°N to 35°S.

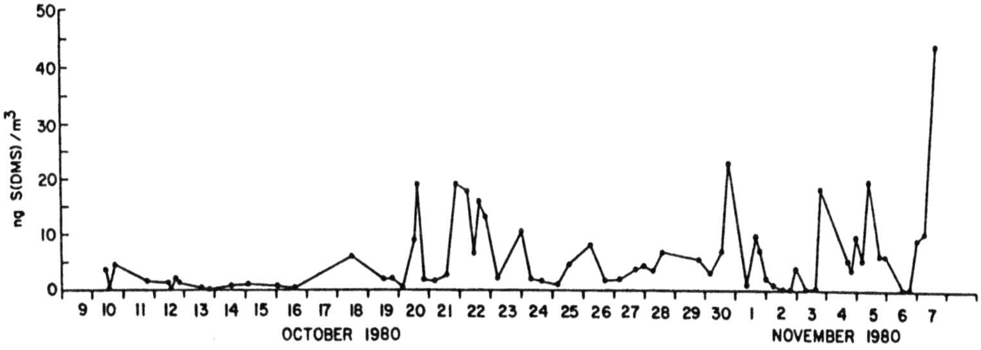

Fig. 4 Concentration of DMS along the cruise track of RV METEOR

The DMS concentrations obtained in this study compare well with those obtained by Andreae et al.(1981) in the eastern Gulf of Mexico, where a mean of 9.7 ng S(DMS)/m^3 was obtained from 36 samples over the open sea. Nguyen et al.(1980) have determined DMS in marine air by an independent method; they have found an average concentration of 3 and 6.5 ng S(DMS)/m^3 over the Mediterranean Sea and the Atlantic Ocean, respectively.

A pronounced increase of atmospheric DMS is evident around October 19, 1980. This coincides with the transition from the zone of prevailing westerlies into the trade wind system, as shown by isobaric airmass trajectory analysis (V.Johannes, 1981, pers. comm.). These airmasses of the trade wind system had been over the ocean for about 2 to 3 days since leaving the african continent, according to the trajectory studies.

Aerosol samples collected during the same time show clear evidence of the advection of dust from the Sahara region (Andreae, unpublished data). The interpretation of this increase in DMS concentrations remains quite ambiguous. The residence time of DMS in the atmosphere is too short to expect an origin of the DMS by advection from the African continent. The Cape Verde Islands were upwind of the Meteor on October 22, 1980; reefs surrounding these islands are a potential source of DMS (Andreae,1981). A similar increase in atmospheric DMS was observed during a more recent cruise (Meteor 57/I, Bingemer, unpublished data), when the Meteor passed the island of Madeira. A few days later, on October 26, airmasses originating from the Amazon were sampled. These samples did not show levels of DMS comparable to those found in the African airmasses. This is a further argument against a terrestrial source for the elevated levels of DMS found in the northern trades region, as the densely vegetated Amazon region should be a much stronger source of biogenic sulfur than the arid regions of North Africa. While considerable fluctuations occur in the atmospheric DMS levels measured along the cruise track south of 25°N all the way to Montevideo, it is still evident from Figure 4 that the average DMS levels south of 25°N were considerably higher than those observed north of this parallel: 7.o vs. 2.5 ng S(DMS)/m^3. A similar difference is not observed in the seawater concentrations (Barnard et al., 1981); therefore the increase in atmospheric DMS south of 25°N cannot be explained by an increase in source strength from the ocean surface. During the last part of the cruise, about from October 27 until the arrival in Montevideo, the Meteor remained relatively close to the South American continent (up to 300 km offshore). Onshore winds prevailed during most of this period, and no systematic changes in DMS levels from those observed on the open ocean were observed until we entered the estuary of the Rio de la Plata, where the highest levels of DMS during this cruise were observed both in the water and in the atmosphere.

With the exception of the Rio de la Plata, no correlation is evident between the concentrations of DMS in air and water. This suggests that variations in the source strength are not responsible for the variations in atmospheric DMS. More likely, these variations are controlled by turbulent mixing in the marine boundary layer and by parameters which control the rate of conversion of DMS, e.g. solar radiation, OH radical concentration, and possibly aerosols which may adsorb DMS or catalyze its oxidation.

3.2. DIURNAL VARIATION OF DMS IN THE MARINE ATMOSPHERE

The reaction of DMS with photochemically generated OH radicals is believed to be the dominant sink process for atmospheric DMS. As a consequence, photochemical models of the marine atmosphere (Graedel, 1979) predict a pronounced diurnal variation of the DMS concentration, showing an inverse relationship to the solar radiation flux and the concentration of OH. Graedel (1979) arrives at a midday minimum of ca. 3 ng S(DMS)/m^3 and a continuous increase during the night up to levels of about 70 ng S(DMS)/m^3 at around 0500 h. The data from the Meteor expedition (Fig. 5) do not show a significant diurnal variation, in agreement with the observations of Andreae et al.(1981). The predicted noon minimum of 3 ng S(DMS)/m^3 approximates well the midday average from our data but the nighttime maximum is much in excess of the observations.

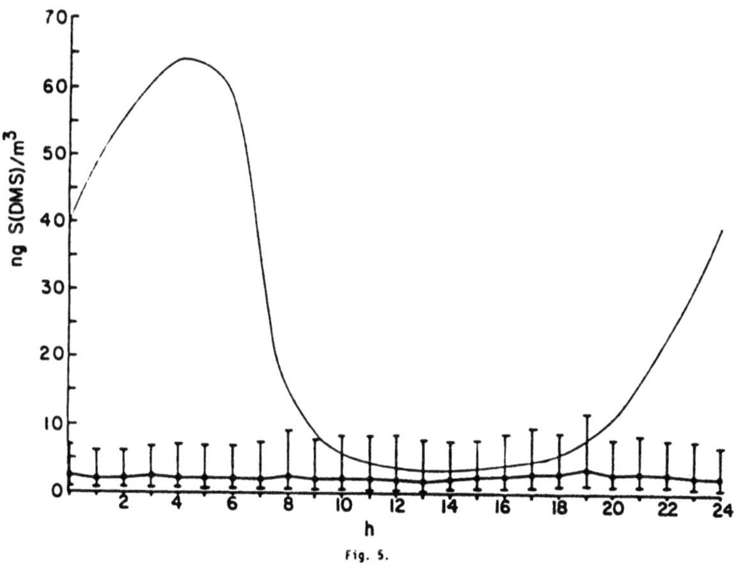

Fig. 5 Diurnal variation of DMS as predicted (Graedel,1979) (thin line) and as observed (vertical bars) during Meteor 56/I. Observations are represented as geometric means and 1σ confidence intervals for each hour.

A number of reasons could explain these discrepancies:

I. Systematic analytical errors could cause the measurements to be too low for the atmosphere and too high for the seawater, resulting in an overestimate of the sea-to-air flux. This appears unlikely, since the agreement of the data from different laboratories (Andreae et al.,1981; Nguyen et al.,1980; Nguyen et al.,1978), obtained with independent analytical methods and different standards, seems reasonable.

II. The flux of DMS from the sea, serving as input in the model calculations for the atmosphere, may be less than calculated on the basis of gas transfer models, reflecting the uncertainties in our understanding of the physics of the gas exchange process, the chemistry of DMS in sea-water, and the accuracy of the constants used in the calculations.

III. There may be significant atmospheric sink processes for DMS which have not yet been identified. The absence of a nightly increase of DMS levels to the extend predicted by the photochemical models suggests that these sinks may not be photochemical, or that they do not consist of reactions with short-lived products of photochemical reactions.

3.3. THE FLUX OF DMS FROM THE OCEAN TO THE ATMOSPHERE

A stagnant film model (Liss and Slater,1974) was used to predict the sea-to-air flux of DMS from the data of M.O. Andreae on DMS in seawater and from the data on DMS in air. In this model the flux between liquid and gas phase is determined by the rate of molecular diffusion across a stagnant film with the thickness z. The concentrations in the gas and liquid phase are c_g and c_l, respectively. The transfer rate is determined by the exchange constant k:

$$F = k\left(\frac{c_g}{H} - c_l\right) = \frac{D}{z}\left(\frac{c_g}{H} - c_l\right)$$

(H: Henry's law constant; D: coefficient of molecular diffusion of gas in stagnant layer.)

In the case of DMS, the atmosphere is highly undersaturated with respect to the coexisting seawater ($H = 0.3(g/cm^3$ air)/(g per cm^3 water), (Lovelock et al.,1972)), so that

$$\frac{c_g}{H} \ll c_l \qquad \text{and thus} \qquad F \cong - k \cdot c_l$$

Different approaches have been used to obtain the exchange constant k:

a) Liss and Slater give a mean value of 19.2 cm/h for DMS.

b) Based on wind tunnel experiments, Kanwisher (1963) proposes a quadratic dependence of k on the wind velocity v:

$$k \sim v^2$$

c) Based on field experiments with the radon method, Kromer (1979) gives a bilinear function for the dependence of k on v:

$$k = 0.075\ v \qquad \text{for } 0 < v < 4 \text{ m/sec}$$
$$k = 0.72(v-4) + 0.3 \text{ for } v > 4 \text{ m/sec}$$

The DMS concentrations and windspeeds measured during the Meteor 56/I cruise were substituted into the model, and average fluxes for the region traversed were computed.
The following results were obtained:

a) 1.04 ng S(DMS)/cm^2h

b) 0.80 ng S(DMS)/cm^2h

c) 0.63 ng S(DMS)/cm^2h

Assuming that these data are representative for the World ocean, the following global sea-to-air fluxes were estimated:

a) 33 Tg S/yr

b) 25 Tg S/yr

c) 20 Tg S/yr

4. CONCLUSIONS

Dimethylsulfide from marine biological sources is the dominant reduced sulfur gas dissolved in seawater. The open ocean serves as a source of gaseous sulfur for the atmosphere by continuous widespread emission of dimethylsulfide.

The source strength of the ocean, as estimated from the data on DMS in the surfacewater by a stagnant film model, falls in the low range of the fluxes predicted by the mass-balance models for the global sulfur cycle. The fluxes calculated by the film model agree fairly well with the estimate of Nguyen et al.(1978), who gives a flux of 27 Tg S/yr on the basis of his seawater measurements.

Considerable discrepancies still exist between the concentrations of DMS in the marine atmosphere predicted by photochemical considerations and the concentrations measured over the ocean. This may be due to uncertainties in the source estimates which serve as input for the photochemical models as well as to a lack in understanding the sink mechanism of DMS in the air.

5. REFERENCES

AMMONS, J.M., Preconcentration methods for the determination of gaseous sulfur compounds in air, Ph.D. Dissertation, University of South Florida, Tampa, Florida, 1980

ANDREAE, M.O., W.R. BARNARD and J.M. AMMONS, The biological production of dimethylsulfide in the ocean and its role in the global atmospheric sulfur budget, Proc. 5th Intern. Symposium on Environmental Biogeochemistry, Stockholm, Sweden, June 1-5, 1981

ATKINSON, R., R.A. PERRY, and J.N. PITTS, Jr., Rate constants for the reaction of OH-radicals with COS, CS_2, and CH_3SCH_3 over the temperature range 299-430 k, Chem. Phys. Lett., 54, 14-18, 1978

BARNARD, W.R., M.O. ANDREAE, W.E. WATKINS et al., The flux of dimethylsulfide from the oceans to the atmosphere, paper presented at I.A.M.A.P. Symposium on Role of Oceans in Atmospheric Chemistry, Hamburg, August 1981; submitted to J. Geophys. Res.

BRAMAN, R.S., J.M. AMMONS, and J.L. BRICKER, Preconcentration and determination of hydrogen sulfide in air by flame photometric detection, Anal. Chem., 50, 992-996, 1978

BREMNER, J.M., and STEELE, C.G., Role of microorganisms in the atmospheric sulfur cycle, Advances in Microbial Ecology, 2, 1978

CHALLENGER, F., Biological methylation, Adv. Enzymology, 12, 429-491, 1951
COX, R.A., and D. SHEPPARD, Reactions of OH radicals with gaseous sulfur compounds, Nature, 284, 330-331, 1980
ERIKSSON,E., The yearly circulation of sulfur in nature, J. Geophys. Res., 68, 4001-4008, 1963
GRAEDEL, T.E., The homogeneous chemistry of atmospheric sulfur, Rev. Geophys. Space Phys., 15, 421-428, 1977
GRAEDEL, T.E., Reduced sulfur emissions from the open oceans, Geophys. Res. Lett., 6, 329-331, 1979
GRANAT, L., H.RHODE, and R.O. HALLBERG, The global sulfur cycle, Ecol. Bull., 22, 89-134, 1976
JANSSEN-SCHMIDT, Th., E.P. RÖTH, et al., Anthropogene Anteile am atmosphärischen Schwefel- und Stickstoffkreislauf und mögliche globale Auswirkungen auf chemische Umsetzungen in der Atmosphäre, Berichte der Kernforschungsanlage Jülich, Nr. 1722, 1981
KURYLO, M.J., Flash photolysis resonance fluorescence investigation of the reaction of OH radicals with dimethylsulfide, Chem. Phys. Lett., 58, 133-137, 1978
KANWISHER, J., On the exchange of gases between the atmosphere and the sea, Deep Sea Res., 10, 195-207, 1963
KROMER, B., Gasaustausch zwischen Atmosphäre und Ozean - Feldmessungen mit der Radonmethode, Ph. D. Thesis, Universität Heidelberg, 1979
LISS, P.S., and P.G. SLATER, Flux of gases across the air-sea interface, Nature, 247, 181-184, 1974
LOVELOCK, J.E., R.J. MAGGS, and R.A. RASMUSSEN, Atmospheric dimethylsulfide and the natural sulphur cycle, Nature, 237, 452-453
NGUYEN, B.C., A. GAUDRY, et al., Reavaluation of the role of dimethylsulfide in the sulphur budget, Nature, 257, 637-639, 1978
NGUYEN, B.C., B. BONSANG, et al., Rôle de la source océanique des composés soufres gaseux dans le cycle atmospherique du soufre, paper presented at the conference on physicochemical behavior of atmospheric pollutants, Berlin, Germany, May 6-7, 1980

On a Diurnal Variation of NO_2 Found in North Atlantic Air Masses

Günter Helas

Max-Planck-Institut für Chemie (Otto-Hahn-Institut), Mainz, FRG

S u m m a r y

In pure marine air masses measured at the west coast of Ireland mixing ratios of NO_2 were found to be lower than 100 pptv and mixing ratios of NO were lower than 10 pptv most of the time. In addition a diurnal variation in the mixing ratios of NO_2 was found with higher daytime values than at night, 84 ± 49 pptv NO_2 overall and 37 ± 6 pptv NO_2 at night.

1. Introduction

During measurements of NO_x carried out at the west coast of Ireland in June 1979, a diurnal variation of NO_2 was found when clean marine air was sampled. It is the purpose of this paper to describe and discuss this diurnal variation.

2. Measurement technique and measurement site

The instrument developed for field measurements has been described previously (Helas et al.,1981). Basically it is similar to commercial NO_x analyzers, which determine the NO level by the chemiluminescence resulting from the nitric oxide-ozone reaction. A number of design changes have been made to improve the sensitivity so as to enable the detection of NO_x mixing ratios in the pptv range. NO_2 was reduced to NO by means of two converters; one is ferrous sulfate kept at ambient temperature, the other is molybdenum chips heated to ~680 K. Both were proven in the laboratory to have reduction efficiencies of close to one hundred percent, but in the field the hot molybdenum often was found to give higher signals than $FeSO_4$. This indicates that nitrogen containing substances other than NO_2 are present in the atmosphere and are at least partially reduced to NO. Here only the results obtained with the $FeSO_4$ converter will be discussed. The instrument was provided with a switches device so that the air being sampled entered consecutively (i) directly to measure NO, (ii) via ferrous sulfate to determine NO_2 + NO, and (iii) via the molybdenum converter to obtain the signal for NO and total convertable compounds. The sampling was done in this sequence, so that each measuring mode takes about 20 minutes during each hour. The first four minutes of one mode are discarded in order to let the instrument adjust to the new conditions, the other 16 minutes are taken to be representative for the entire hour, an approximation which appears permissible for background air. The instrument was calibrated once or twice daily. The procedure made use of a reference gas, which was checked before and after the measurement campaign. The zero level was determined by oxidizing the NO of the sample air to NO_2 with ozone before it entered the reaction chamber. The detection limit was found subject to minor changes in sensitivity and was normally about 5 pptv, but could occasionally go up to 10 pptv. This latter value is chosen as the more conservative detection limit.

The measurements were made during the time period of June 12 - 27, 1979. The measurement site was on the grounds of the Loop Head light house at the tip of the Loop Head peninsula, the northern estuary region of the Shannon river in Ireland. The sampling site is approximately 70 m above sea level and is open to the North Atlantic Ocean in south-western, western, and north-western direction.

3. Results

The average mixing ratio of NO_2 for the entire period was 0.101 ± 0.087 ppbv. The high variability is to be noted. It must be ascribed at least in part to the influence of continental sources. Two periods were encountered during which pure marine air masses arrived at Loop Head as confirmed by trajectory analysis. The present communication will concentrate on these latter data. In pure marine air masses, the NO mixing ratios encountered often were very low, so that they were in the noise band of our instrument, i. e. lower than 10 pptv.
Fig. 1 shows, as an example, the approach to one of the two periods of pure marine air. Mixing ratios of NO_x are plotted for the time of June 13,0h to June 15, 24h. The crosses denote NO_2, the open circles NO. As mentioned above, the molybdenum converter reduces in addition to NO_2 other N-containing compounds to NO too. Until their nature is satisfactorally described, our working term for these additional compounds is excess-NO_x. Their amount is given by the full dots.

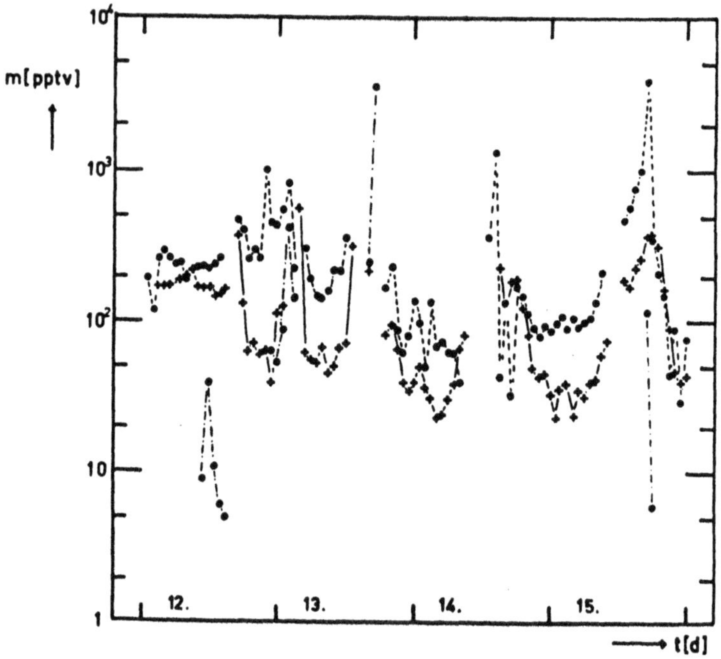

Fig. 1 Mixing ratios of NO_2 (crosses), NO (open circles), and excess-NO_x, as explained in the text (full dots), for the time of June, 12 to June, 15

During the time of measurements shown in fig. 1, there was a shift in wind direction and the back-track calculations (Schütz 1980) showed a distinct change from air masses having touched the continent to air masses of maritime origin, as is shown in fig. 2 a-d. Here the arrow indicates the location of Loop Head in Ireland. The crosses refer to surface trajectories, the open circles to 850 mbar trajectories, which approximate a level of 1500 m height and thus are less subjected to orographic disturbances. The trajectory calculations were made for 48 hours, the intervals

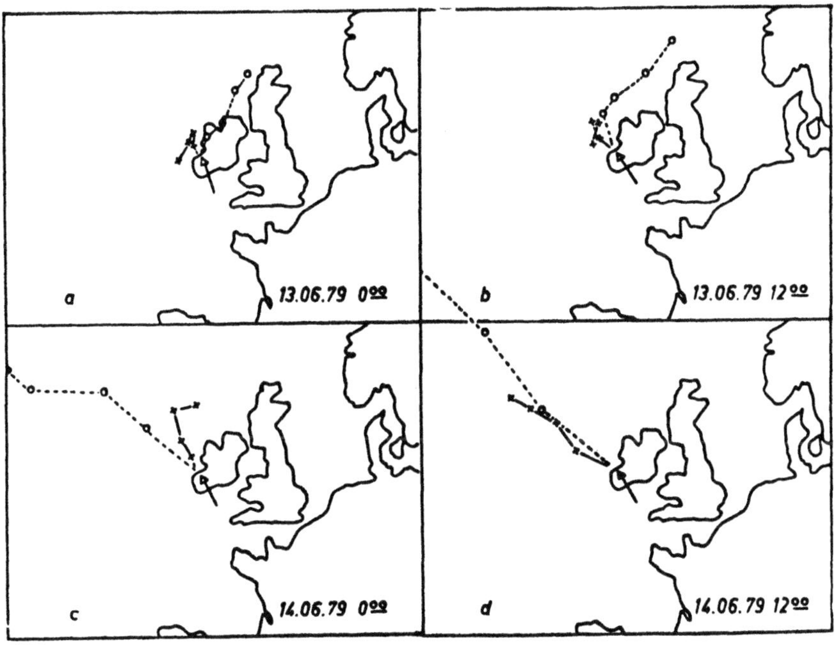

Fig. 2 Back-trajectories for air masses arriving in Loop Head on June 13, and 14. The arrow indicates the position of Loop Head, the crosses and solid lines refer to surface level, and open circles and dashed lines to 850 mbar level.

used 12 hours. The sequence a to d in fig. 2 shows the transition from a slow movement of air masses due to influence of a high pressure region to a fast steady inflow of marine air masses. This change of origin of air masses corresponds to the decrease of NO_x mixing ratios during June 13 - 14, as is shown in fig. 1.

In pure marine air, which was encountered from June 14, 1h to June 16, 5h and June 20, 4h to June 23, 6h, the average NO_2 mixing ratio was 84 ± 49 pptv. Again, the variability is high. In this case it was found due to a diurnal variation. The data shown in fig. 3 were obtained by averaging hourly data. The diurnal variation there is clearly displayed.

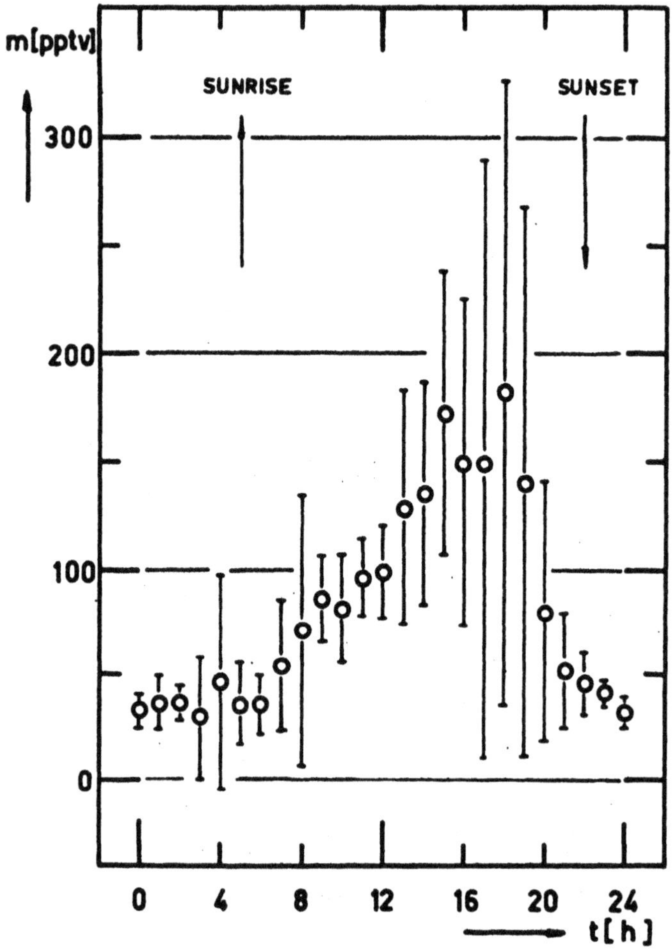

Fig. 3 Mixing ratios of NO_2 in marine air masses as a function of local time

The night-time values remain fairly constant as indicated by the small standard deviation of 6 pptv. The higher values are correlated with the daylight, the time of sunrise and of sunset is indicated by the arrows.

4. Discussion

The strict dependency of increase and decrease of NO_2 mixing ratios on daylight points to a photoactive source. Suitable molecules in the gas-phase like N_2O_5, HO_2NO_2, $CH_3O_2NO_2$, PAN, which could act as precursors for NO_2 have much smaller absorption cross sections than NO_2 itself (Cox and Tyndall,1979; Graham et al.,1978; Johnston and Graham,1974). So they are unlikely to act as a source for NO_2.
The thermal decay of precursor compounds formed during the night would require sufficient temperature changes during the day. The air temperature was 13 ± 2 °C at the measurement site during the sampling campaign, which rules the thermal effects out. The low mixing ratios of NO_2 of about 35 pptv at the surface indicate a downward flux of NO_x as the lower stratosphere mixing ratios are known to be approximately 300 pptv (Blatherwick et al.,1980). An enhanced vertical downward mixing during the day from the colder upper troposphere is not likely, as rather cold air masses were encountered which were travelling over a warm ocean surface leading to a continuous vertical exchange during both day and night.
As it is difficult to explain the formation of NO_2 by gas-phase processes, a production of NO_2 or precursors in the liquid phase should be considered. Zafiriou et al. have demonstrated, that NO is formed during photolysis of nitrite contained in seawater (Zafiriou and True,1979; Zafiriou et al.,1980). This could be a possible process, but unfortunately we have no measurements of evolution of NO or NO_2 in the off-shore area. It is also conceivable, that similar photolytic processes take place in the liquid phase of the aerosol, but this is clearly speculative at present.
The time constant of decline of NO_2 mixing ratios at the

evening is about 3 hours. If NO_2 arises from an off-shore marine source, it is possible that the decline then is due to the diminishing of source strength and the sweep out of NO_2 by advection of air with lower mixing ratios. Such an explanation would not hold,however, if the source were an areal source covering a larger part of the ocean. In this case the decline of NO_2 would indicate a true atmospheric lifetime of ~3 hours.
Two chemical reactions are known, which may act as a sink for NO_2. The caracteristic lifetime for NO_2 according to the reaction

$$NO_2 + O_3 \longrightarrow NO_3 + O_2$$

with 30 ppbv ozone, an average that is typical for the location, is approximately 11 hours(Becker et al.,1974).
During the day NO_3 is reverted back to NO_2 by photolysis and by thermal decay.
The main sink for NO_2 is the reaction with the OH radical.

$$NO_2 + OH \longrightarrow HNO_3$$

The molecule formed is HNO_3 and the product is removed by wet deposition. The characteristic lifetime for NO_2 for an average number density of 10^6 cm^{-3} (Crutzen and Fishman,1977) of OH is in the range of one day (Zellner,1978). This reaction is also slower than the decline found. So the rapid decline could only be understood as a replacement of source modulated air, or one must look for another, yet unknown sink.

The NO_2 mixing ratios, we have observed in Atlantic air masses, may be used to estimate the residence time of NO_2. We can derive a travel time of seven days, if we assume the air masses were originated from the North American continent. At the populated east coast of United States and Canada, the NO_2 mixing ratios will be about 10 ppbv and our night-time average in Atlantic air masses is 35 pptv. If one uses the expression $m = m_o\ (-t/\tau)$, we calculate a residence time of $\tau = 1.2$ days. Here it is assumed that the ocean does not act as a source, so it must be considered as an upper limit. A lower limit to the residence time is indicated by the diurnal cycle. If this rapid decline is due to a real loss of NO_2 and not due to source variation, the atmospheric residence time would be three hours. The evaluated residence times of 1.2 days or 3 hours are either the limitation for a value, which has to be better established, or points to two different loss processes of NO_2 in the atmosphere.

5. Acknowledgements

We thank the Commissioners of Irish Lights, especially H. N. Greenlee, for the possibility to use their facilities at the Loop Head light house for these measurements.

6. References

Becker, K. H., Schurath, U., and Seitz, H.,1974
Ozone-olefin reactions in the gas phase 1. Rate constants and activation energies, Int. J. Chem. Kinet. 6,725-739, 1974

Blatherwick, R. D., Goldman, A., Murcray, P. G., Murcray, F. J., Cook, G. R., and VanAllen, J. W.,1980
Simultaneous mixing ratio profiles of stratospheric NO and NO_2 as derived from balloon-borne infrared solar spectra Geophys. Res. Lett. 7,471-473,1980

Cox, R. A. and Tyndall, G. S., 1979
Rate constants for reactions of CH_3O_2 in the gas phase
Chem. Phys. Lett. 65,357-360,1979

Crutzen, P. J. and Fishman, J., 1977
Average concentrations of OH in the troposphere and the budgets of CH_4, CO, H_2, and CH_3CCl_3
Geophys. Res. Lett. 4,321-324,1977

Graham, R. A., Winer, A. M., and Pitts, J. N., 1978
Ultraviolet and infrared absorption cross sections of gas phase HO_2NO_2
Geophys. Res. Lett. 5,909-911,1978

Helas, G., Flanz, M., and Warneck, P., 1981
Improved NO_x monitor for measurements in tropospheric clean air regions
Int. J. Environ. Anal. Chem. 10,155-166,1981

Johnston, H. S. and Graham, R., 1974
Photochemistry of NO_x and HNO_x compounds
Can. J. Chem. 52,1415-1423,1974

Schütz, L., 1980
the program for the trajectory calculations was kindly supplied by L. Schütz

Zafiriou, O. C. and True, M. B., 1979
Nitrite photolysis as a source of free radicals in productive surface waters
Geophys. Res. Lett. 6,81-84,1979

Zafiriou, O. C., McFarland, M., and Bromund, R. H.,1980
Nitric oxide in sea water
Science 207,637-639,1980

Zellner, R., 1978
Recombination reactions in atmospheric chemistry
Ber. Bunsenges. Phys. Chem. 82,1172-1179, 1978

A Discussion of Approximate Relations for Transfer and Deposition of Trace Constituents in the ABL*

F. Herbert und G. Kramm

Institute of Meteorology and Geophysics, J. W. Goethe-University Frankfurt/Main, FRG

S u m m a r y : Calculation methods of the vertical diffusivity coefficients, the transport and the deposition velocity of a trace constituent in the ABL are presented. For determing actual values and profiles these methods are applicable to field measurements obtained from instrumented towers or boundary layer soundings. The theoretical concept is general enough to enable numerical evaluations at variable stability conditions in the atmosphere.

1. INTRODUCTION

A number of atmospheric features, thermal stability, baroclinicity, the momentum of upper level winds, topographic and surface layer characteristics as well as transfer resistances combine together to predict the turbulence parameters which are responsible for transport, dispersion and deposition of trace constituents in the ABL. As environmental problems by air pollution increase it becomes gradually important to measure and model these effects. Their contribution to the mass budget of a trace constituent is determined by the turbulent vertical flux divergence. The interactions between the transport fields and the necessity of available field observations cause considerable difficulties when numerical modeling is attempted for turbulent transports.

To obtain tractable expressions for the transfer of a trace constituent it is advantageous to link it to the transfer of water vapor because one may assume that trace constituents and vapor follow quite similar transport mechanisms in the ABL. This similarity might be natural. Furthermore, it is reasonable to adopt similarity between humidity and buoyant

---

* Part of this paper was presented in advance by G. Kramm at a separate SFB-colloquium in june 1981 at the University of Mainz

heat transport. This similarity was recommended for practical calculations by Delsol et al. (1971) and its evidence has been confirmed by investigations of Swinbank and Dyer (1967), Högström (1967) and Dyer and Hicks (1970).

The prime objecitve of the study of the transfer of a trace constituent is to determine its mass flux within the surface ABL and, in particular, which is transferred through this layer down to the surface. The surface layer is the lowest part of the ABL close to the earth surface. In this layer the interaction of air with the (earth-)surface is so strong and adjustment to the surface conditions passes off so rapid that stationary conditions may be assumed within the surface ABL. A quasi-steady state can relatively good be supposed for air in the lowest 20-50 m above the surface where, in addition, the fluxes of momentum and heat are usually considered as constant, i.e. independent of height.

2. THE FLUX PROBLEM

The complex behavior of turbulent transfer in the ABL is such that semi-empirical relationships must be developed which use eddy diffusivity (transfer)coefficients K to define vertical flux-gradient equations. According to this method, the mass flux F_q of a trace constituent with mean concentration q (= mass fraction) is

$$F_q = - \rho_0 \; K_h \frac{\partial q}{\partial z} \tag{1}$$

where z is the height, ρ_0 is a reference density of air, and K_h is the eddy heat diffusivity. Note in (1) that we introduced K_h as transfer coefficient because of the evidence of similarity between turbulence transfer of trace constituents, vapor and buoyant heat. A second method to describe F_q is given using the eddy diffusivity in integrated form. This treatment is to evade complicated gradients $\partial q/\partial z$ for practical applications and to determine the eddy flux proportional to a finite difference Δq of the concentration of a trace constituent (or any other charcteristic property) to be transferred from the surface upward to the height or viceversa. Although this method leads to principal and practical difficulties in applying it to greater heights or even to the total ABL, nevertheless, it might be useful for constant or quasi-constant flux conditions in the surface ABL. Then, for the appraoch F_q = const. it is equivalent to use instead of (1) the integrated flux equation

$$F_q = - \frac{\rho_0 \ (q_2 - q_1)}{\int_{z_1}^{z_2} K_h^{-1} \, dz} \qquad (2)$$

where q_1 and q_2 are concentration values measured on two levels z_1 and z_2 within the surface layer. The integral expression

$$\{ \int_{z_1}^{z_2} K_h^{-1} \, dz \}^{-1} = r^{-1}$$

in (2) then plays the role of a 'bulk' conductivity (resp. a resistance r) with respect to the interval $\Delta z = z_2 - z_1$. Note that, setting the lower boundary equal with the surface, i.e. $z_1 = z_0$ or 0, and extending the upper boundary to the ambient environment ($z_2 \rightarrow \infty$) it theoretically defines a total bulk conductivity (or resistance) coefficient.

In the case we want to study the transfer rates of trace constituents in the ABL it is desirable to evaluate one of the expressions given in (1) or (2). Both, the differential gradient formulation (1) and the integral bulk formulation (2) primarily depend on the knowledge of the eddy heat diffusivity K_h which must be determined as function of the mechanical and thermal state caused by the turbulence activity in the ABL. The complexity of this problem is evident. Especially, in order to describe the complex turbulent interactions in the surface layer under constant flux conditions it is customary to use height independent scaling factors for wind speed (u_* = friction velocity), temperature (Θ_*), moisture (m_*) and trace constituent (q_*). Instead of the original fluxes we then may determine the four unknowns u_*, Θ_*, m_* and q_* by the following set of equations

$$\left.\begin{aligned} u_*^{\,2} &= \frac{u_2 - u_1}{\int_{z_1}^{z_2} K_m^{-1}\,dz}, \qquad \Theta_* = u_*^{-1}\,\frac{\Theta_2 - \Theta_1}{\int_{z_1}^{z_2} K_h^{-1}\,dz} \\ m_* &= \frac{(m_2 - m_1)\Theta_*}{\Theta_2 - \Theta_1}, \qquad q_* = \frac{(q_2 - q_1)\Theta_*}{\Theta_2 - \Theta_1} = \frac{(q_2 - q_1)m_*}{m_2 - m_1} \end{aligned}\right\} \quad (3)$$

wherein similarity between the q and the temperature and humidity profile is considered. The general constituent flux (2) then is equivalently determined by

$$F_q = -\rho_0\, u_*\, q_* = \frac{F_m\,(q_2 - q_1)}{m_2 - m_1} \qquad (4)$$

using the knowledge of the vapor flux F_m. The constant flux relations indicate that to obtain the scaling parameters and the fluxes, it is necessary to have observations of all variable parameters in two heights within the surface ABL and to make assumptions about the eddy diffusivities of heat K_h and momentum K_m. The profile structure of these transfer coefficients is subject of the next section.

A note to the practical application of (1) and (2) seems to be appropriate here. Since the actual state of turbulence is exclusively responsible for the transfer of a trace constituent a principal and decisive problem to determine F_q is to have significant observations of the profiles of wind, temperature and the composition of air. Moreover, in the case that (1) is applied to find F_q , the situation is rather impeded since F_q directly depends on the actual vertical gradients of these parameters which, of course, can hardly be measured with sufficient accuracy. Therefore, evaluation of (1) for determining trace constituent fluxes with the aid of gradient measurements is rather questionable, unless we are willing to accept arbitrarily large truncation errors in the prediction of F_q. Quite different appears the situation in the case when the integrated formulation can be assumed to predict transfer rates in the constant flux surface layer. In order to determine now F_q from (2) it is usually sufficient as quoted above to make observations of wind speed, temperature and air composition at two characteristic heights z_1 and z_2. Practical experiences show that measurements of these variables in lower heights are in any case extremely more reliable than measurements of their vertical gradients. On the other hand we may not ignore the theoretical advantage of the original flux-gradient formulation (1) for greater heights within the ABL since the integrated method (2) can merely be relevant in air under constant flux conditions near to the earth's surface.

3. EDDY DIFFUSIVITIES

General relations: Of particular importance in modeling the ABL is the specification of significant profile expressions for the eddy diffusivities K_m and K_h which depend on internal and external parameters. One of the main difficulties thereby is to predict K_m and K_h as functions of thermal stability of the air, i.e. as functions of the Richardson-number

$$Ri = \frac{g}{\Theta_m}\left[\frac{\partial\theta}{\partial z} + \theta_m\ (R_v/R_a-1)\ \frac{\partial m}{\partial z}\right]\left(\frac{\partial\vec{u}}{\partial z}\right)^{-2} \tag{5}$$

All parameters in (5) have standard notations where θ_m is a characteristic temperature and R_v/R_a-1 is the vapor buoyancy ratio $\simeq 0.61$. Appropriate equations for the eddy diffusivities can be derived in the context with the general concept of the mechanical and thermal ABL-energy transfer rates. With this method argued in Herbert (1979) we can find the following general coefficient profile equations:

$$K_m = l^2\left|\frac{\partial\vec{u}}{\partial z}\right|\ (1-Ri)^{1/2}\quad ;\qquad Ri \leqslant 1 \tag{6}$$

in stable and unstable air, as well as

$$K_h = K_m\quad ;\qquad 0 \leqslant Ri \leqslant 1 \tag{7}$$

in stable and neutral air, and

$$K_h = (\kappa z)^{-1}\left|\frac{\partial\vec{u}}{\partial z}\right|^{-1/2} K_m^{3/2}\quad ;\qquad -\infty < Ri < 0 \tag{8}$$

in unstable air. The factor κ is the von Kármán constant and is approximately equal to 0.4 while l is the so-called mixing

length, a function of height which strongly depends on the actual conditions of the ABL. The formulation of l as function of z is usually based on empirical considerations.
In the present study we use Blackadar's (1962) l-approach

$$l = \frac{\kappa z}{1 + \frac{\kappa z}{\lambda}} \quad , \quad \text{with } \lambda = \text{fkt (Ri)} . \tag{9}$$

Eq. (9) interpolates l between Prandtl's approach $l = \kappa z$ for very low levels close to the surface ($z \to 0$) and the maximum value $l = \lambda$ reached at the upper levels of the ABL; the asymptotic value λ is related to external and internal parameters. In particular, (9) turns out to be quite useful to explore eddy diffusivity profiles when λ can be set equal to a reasonable,constant value. In this study we take for λ the formula

$$\lambda = 0.009 \; u_*/f \tag{10}$$

suggested by Fiedler (1972) in analogy to the height parameter in Lettau's (1962) l-formulation.
Examples of numerical calculations of K_h and K_m-profiles are plotted in Figure 1 and 2. To abtain these curves we evaluated (6) to (8) and (9) with smoothed wind and stratification profiles obtained from ABL-soundings over the city of Köln at stable and near-neutral conditions. The λ-values were defined with (10) where the value for u_* was computed in the lowest level. For comparison of the computed K_m and K_h-profiles note the extremely different vertical distributions of the coefficients provided with the use of the stability functions S_m and S_h found by Businger et al. (1971).

<u>Approximate relations:</u> In the case we want to study the transfer of a trace constituent in the lower ABL it is advantageous to adjust the general eddy diffusivity expressions to surface layer conditions. Two extremely different cases shall be briefly discussed. First we explore the situation when we may use the linear l-approach $l = \kappa z$ as approximation for the mixing length in the lowest few meters above

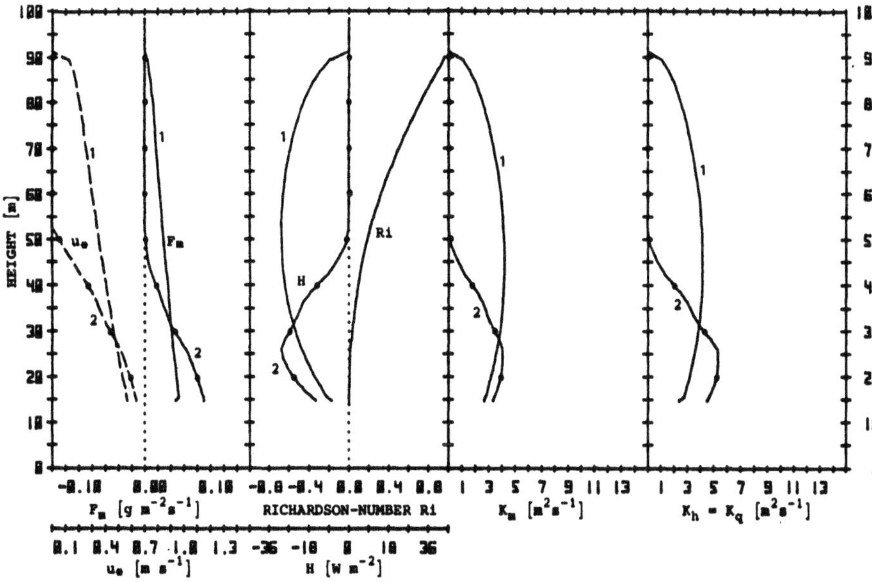

Figure 1: Stable air profiles of the friction velocity u_*, the turbulent fluxes of heat (H) and humidity (F_m) and the eddy diffusivities K_m and K_h as functions of Ri-numbers.
Case 1: Application of Eqs. (6) and (7).
Case 2: Application of Businger's et al. (1971) stability functions S_m and S_h.

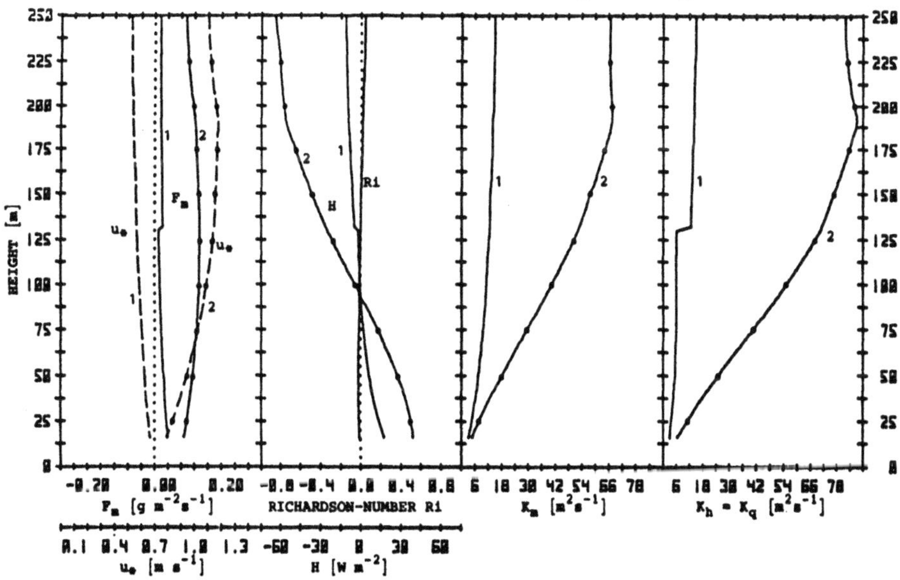

Figure 2: Profiles of the friction velocity u_*, the turbulent fluxes of heat (H) and humidity (F_m) and the eddy diffusivities K_m and K_h as functions of Ri-numbers in near-neutral (unstable-stable) air.
Case 1: Application of Eqs. (6), (7) and (8).
Case 2: Application of Businger's et al. (1971) stability functions S_m and S_h.

the surface. In this case the eddy diffusivities are then reduced to the following expressions

Stable and neutral air:

$$K_h = K_m = (\kappa z)^2 \left|\frac{\partial \vec{u}}{\partial z}\right| (1-Ri)^{1/2} \qquad (11)$$

Neutral and unstable air:

$$\left.\begin{aligned} K_m &= (\kappa z)^2 \left|\frac{\partial \vec{u}}{\partial z}\right| (1-Ri)^{1/2} \\ K_h &= (\kappa z)^2 \left|\frac{\partial \vec{u}}{\partial z}\right| (1-Ri)^{3/4} \end{aligned}\right\} \qquad (12)$$

These formulas for K_m and K_h can relatively good be applied to determine profiles of wind, temperature, humidity and trace constituents in the constant flux layer. Of course, this applies especially to air with neutral or weakly stable or unstable stratification (i.e. for Ri-numbers close to the equilibrium value Ri = 0) since (11) to (13) were found with the quasi-neutral mixing length $l = \kappa z$.

A further note of warning when evaluating K_h seems to be appropriate here, because the unstable K_h-relations (8) should be applied with caution. For this let us consider neutral equilibrium, Ri = 0, near to the surface. Although it follows clearly from the general relationships (6) to (8) that the state and neutral K_m-relation should be used to develop the coefficient K_h in neutral air, the neutral relations in the lowest few meters are equivalently provided by the unstable K-relations (11) and (12); the result is the simple Prandtl formula

$$K_h = K_m = K = (\kappa z)^2 \left|\frac{\partial \vec{u}}{\partial z}\right| \quad . \qquad (13)$$

This equivalence is caused by the hypothesis $l = \kappa z$ which is accurate close to the surface but which fails at higher levels. Thus, at greater heights, the neutral air coefficient K_h should be computed from (6) and (7) using a more general l-approach (e.g. Blackadar's formula (9)) but not from the unstable relation (8) which provides wrong results for K_h.

As second case we consider a free convective layer. In free convection we can assume that

$$Ri << -1 \quad . \tag{14}$$

The eddy diffusivities then take the form

$$K_m = l^2 \left|\frac{\partial \vec{u}}{\partial z}\right| \left|Ri\right|^{1/2} \tag{15}$$

$$K_h = \frac{l^3}{\kappa z} \left|\frac{\partial \vec{u}}{\partial z}\right| \left|Ri\right|^{3/4} \tag{16}$$

where l (and the length scale λ if Blackadar's l-formula (9) is employed) are functions of Ri, of course. Apart from the surface layer, it is eminently difficult to derive a closure relation l = l(Ri) valid at arbitrary height in the ABL. Close to the surface a closure condition may be deduced from the familiar $\theta' \sim z^{-4/3}$ similarity law for the temperature gradient Θ' in free convective air (Lumley & Panofsky, 1964). This provides the heat diffusivity coefficient

$$K_h = 0.9 \left|\frac{\partial \vec{u}}{\partial z}\right| \left|Ri\right|^{1/2} z^2 \tag{17}$$

where the pre-factor 0.9 has been suggested by Priestley (1959). Equalizing (17) and (16) yields the mixing length

$$l \approx 0.71 \left|Ri\right|^{-1/12} z \quad . \tag{18}$$

Eq. (18) finally provides the momentum diffusivity coefficient

$$K_m = 0.51 \left|\frac{\partial \vec{u}}{\partial z}\right| \left|Ri\right|^{1/3} z^2 \tag{19}$$

for free convective air in the surface layer.

Remark: More comprehensive investigations of the closure problem give advice how K_m and K_h from (6) - (8) can be predicted at arbitrary Ri-values.Obviously, a quite reasonable and useful way seems to assume that the mixing length parameter is written as a product of a neutral value l_n and a characteristic stability function ϕ(Ri),

$$l = l_n \, \phi(Ri) .$$

The number of empirical and theoretical relations describing the turbulent ABL offer several calculation schemes for ϕ(Ri) which, however, can not be discussed in the scope of this paper. Then, with given function ϕ(Ri), the proper closure problem reduces to specify l_n such as for instance by Blackadar's neutral l-formula (9).

4. DEPOSITION VELOCITY

The eddy diffusivities argued in section 3 are key quantities for the transfer in the ABL. They are therefore extremely responsible for the distribution and the deposition process of a trace constituent which are (about the diffusivities) functions of internal and external parameters. As known, dry deposition is one of the most important effects for the removal of a trace constituent (gas or particles) from the atmosphere; it is usually described by the so-called deposition velocity defined in Chamberlain (1973) or Garland et. al.(1973). In the ABL we may formulate with (2) a general 'turbulent' deposition velocity by the gradient relation (Fig. 3)

$$DV_q = - K_h \, \partial \ln q / \partial z \tag{20}$$

where K_h is considered as known. Of course, a quantitative evaluation of (20) is in the general case extremely complicated because of the same practical difficulties as they arise in (1) to predict F_q in terms of gradient measurements. Nevertheless, a characteristic value of the deposition velocity should be determinable from (20) when we may adopt that DV_q is continously between two heights z_1 and z_2 (Fig. 3)

$$\overline{DV_q} = \frac{\ln (q_1/q_2)}{\int_{z_1}^{z_2} \frac{dz}{K_h}} \tag{21}$$

which represents a mean deposition velocity for the layer $\Delta z = z_2 - z_1$. The integral in this expression can be solved exactly or by approximation for different K_h-approaches. For evaluating (21) it is necessary to have observations at two heights z_1 and z_2 of both the meteorological parameters (wind speed, temperature and humidity) and especially of the trace constituent concentration q. This indicates that, for (21) provides significant values for DV_q in a sub-layer, the measurement levels must be chosen with care. If this is possible, the quantity $\overline{DV_q}^{-1} = R$ may be interpreted as a characteristic resistance of the layer Δz acting on the flux and on the deposition process. By applying this technique to

a number of sub-layers we could get information on the deposition behavior of a trace constituent and the resistance on it by the ambient air in a greater range of the ABL.
At this point we briefly quote a formulation used by Roth (1975) and Lege (1981) which is apparently similar to (21). This expression for DV_q is only applicable in constant flux layer and uses quite another definition which seems not to be appropriate to describe dry deposition of trace constituents on a surface.
In the case we want to know the deposition velocity of a trace constituent near to the ground or even on a land or water surface one possibility is to evaluate (21) under the condition that the two measurement levels z_1 and z_2 are situated close to the earth surface. There we may well assume that momentum, heat and humidity fluxes are independent of height and, due to their condition, that wind speed, temperature and vapor concentration have well-defined, similar vertical profiles being functions of the mechanical-thermal stratification stability. The deposition velocity $\overline{DV_q}$ is then precisely computable for arbitrary stability values by solving the integral in (21). Is it not useful to discuss here the more complicated cases when the atmosphere is in non-neutral conditions. For convenience, we will only explore the situation under neutral conditions. With (13) and the neutral logarithmic wind profile, i.e.

$$K_h = \frac{\kappa^2 (u_2 - u_1)\ z}{\ln\ (z_2/z_1)} \quad , \tag{22}$$

we obtain from (21) as deposition velocity in the surface layer (Fig. 3)

$$\overline{DV_q} = \frac{\kappa^2 (u_2 - u_1)\ \ln(q_1/q_2)}{(\ln\ z_2/z_1)^2} \tag{23}$$

where u_2 and u_1 are the wind speeds at the upper and lower level. Or, when the conditions at the surface are considered as known,

$$\overline{DV_{q_0}} = \frac{\kappa^2 u_1\ \ln(q_0/q_1)}{(\ln\ z_1/z_0)} \tag{24}$$

where z_0 is the surface level where the wind speed vanishes.

Once we recognize that $\overline{DV_q}$ is computable when the concentration at the constituent and the wind speed are available at two heights; in non-neutral air, temperature and humidity must additionally be known. The advantage of this technique is that we don't have to adopt that the constituent flux F_q is constant. On the other hand, the fact that q must be determined at two heights is a drawback since we must expect that (although F_q being not constant) the q-profile has a steep slope in the surface layer and in particular close to the surface. To get reasonable values $\overline{DV_q}$ from (23) or (24) it is probably indispensable that the distance $z_2 - z_1$ or $z_1 - z_0$ is very small which complicates the observational (and technical) procedure expecially as one must take into account that the surface value q_0 may become very small or even zero for several trace constituents.

These difficulties do not occur when F_q is assumed to be constant in the surface layer just as the other fluxes too. In general stability conditions similarity between trace constituent and humidity transfer is reasonable (see (4)). Again, it is convenient to explore the case F_q = const. under neutral conditions where we can suppose (22) as well as similarity of the q and wind profile in virtue of $K_h = K_m$. Then we get from (2) and (3) the following equations for surface flux and concentration of the trace constituent

$$F_q = \frac{-\rho_0\, \kappa^2\, (u_2 - u_1)(q_2 - q_1)}{(\ln z_2/z_1)^2} = \text{const.} \tag{25}$$

$$q_0 = q_2 - \frac{q_2 - q_1}{u_2 - u_1}\, u_2 \tag{26}$$

Hence, the deposition velocity on the surface (Fig. 3)

$$DV_{q_0} \equiv \frac{F_q}{\rho_0 q_0} = \frac{+\,\kappa^2 (u_2 - u_1)^2\, (q_1 - q_2)}{\left\{\ln \frac{z_2}{z_1}\right\}^2 (q_1 u_2 - q_2 u_1)}\,, \tag{27}$$

a formula earlier derived by Businger (1973). Note that the observation heights in (27) may arbitrarily be chosen within the constant flux layer and their distance must not be very small in contrast to the reference heights in (23) or (24). This is the advantage of (27) in comparison to (24); the drawback, however, is the assumption F_q = const. which must not be fulfilled for a trace constituent in the surface layer.

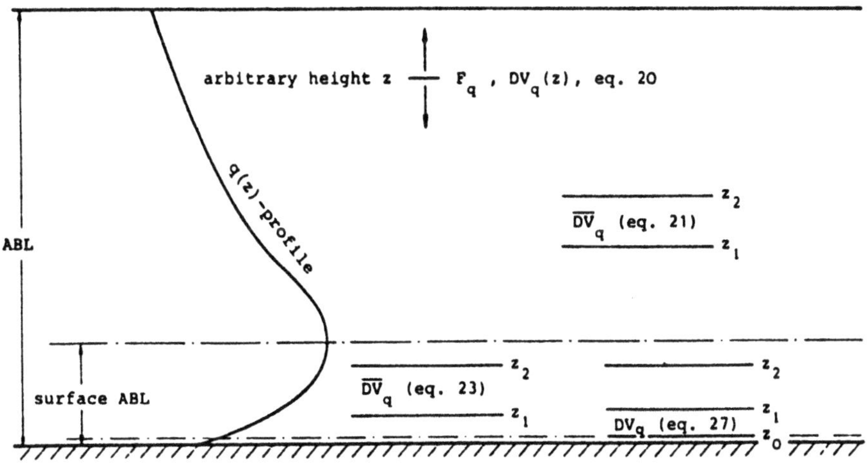

Figure 3: Schematical illustration to the applicability of the various DV_q-formulas

5. FINAL REMARKS

Appropriate semi-empirical expressions can be formulated to describe the eddy transfer of a trace constituent in the ABL over flat terrain. Characteristic expressions to determine vertical profiles of momentum and buoyant heat diffusivity coefficients are presented. They are general enough to be applicable over a wide spectrum of stable and unstable atmospheric states. Their relevance depends on well-adjusted closure conditions for the mixing height scale occuring in the diffusivities. Reliable mixing height relations might be obtained for the total stability regime from empirical similarity laws.

The concept of trace constituent deposition under the turbulent influence of the ABL and its practical application to model significant deposition velocities (resp. resistances) for a sub-layer or at a surface is discussed. The study particularly exhibits that careful and complex field-observations are required in order to can expect from numerical deposition models sufficiently reliable results.

6. REFERENCES

BLACKADAR, A. K., 1962: The vertical distribution of wind and turbulent exchange in a neutral atmosphere. J. Geophys. Res., 67, 3095-3102

BUSINGER, J. A., WYNGAARD, J. C., IZUMI, Y., and BRADLEY, E. F.,1971: Flux-profile relationships in the atmospheric surface layer. J. Atmos. Sci., 28, 181-189

BUSINGER, J. A., 1973: Turbulent transfer in the atmospheric surface layer. In: Workshop on micrometeorology, Haugen, D. A., Ed., Amer. Meteor. Soc., 392 pp.

CHAMBERLAIN, A. C., 1974: Deposition of SO_2 by gaseous diffusion. In: Advances in Geophysics, Vol. 18B, Frenkiel, F. N., and Munn, R. E., Ed., Academic Press, Inc., New York/San Francisco/London, 389 pp.

DELSOL,F., MIYAKODA, K., and CLARKE, R.H., 1974: Parameterized processes in the surface boundary layer of an atmospheric circulation model. Quart. J. R. Met. Soc., 97, 181-208

DYER, A. J., and HICKS, B. B., 1970: Flux-gradient relationships in the constant flux layer. Quart. J. R. Met. Soc., 96, 715-721

FIEDLER, F., 1972: The effect of baroclinicity on the resistance law in a diabatic Ekman layer. Beitr. Phys. Atmosph., 45, 164-173

GARLAND, J. A., CLOUGH, W. S., and FOWLER, D., 1973: Deposition of sulphur dioxide on grass. Nature, 242, 256-257

HERBERT, F., 1979: Energy transfer rates and exchange coefficients in the turbulent atmospheric PBL. Paper presented at the 6th annual EGS-meeting at Vienna, September 1979

HÖGSTRÖM, U., 1967: Turbulent water vapor transfer at difference stability conditions. Phys. of Fluids, 10, Supplement, 247-254

LEGE, D., 1981: Eine Betrachtung der Bestimmung des Stroms fühlbarer Wärme und der Schubspannungsgeschwindigkeit aus Temperatur- und Geschwindigkeitsdifferenzen. Meteorol. Rdsch., 34, 1-4

LETTAU, H. H., 1962: Theoretical wind spirals in the boundary layer of a barotropic atmosphere. Beitr. Phys. Atmosph., 35, 195-212

LUMLEY, J. L., and PANOFSKY, H. A., 1964: The structure of atmospheric turbulence. J. Wiley & Sons, New York/London/ Sydney, 239 pp.

PRIESTLEY, C. H. B., 1959: Turbulent transfer in the lower atmosphere. University of Chicago Press, Chicago, 130 pp.

ROTH, R., 1975: Der vertikale Transport von Luftbeimengungen in der Prandtl-Schicht und die Deposition-Velocity. Meteorol. Rdsch., 28, 65-71

SWINBANK, W. C., and DYER, A. J., 1967: An experimental study in micro-meteorology. Quart. J. R. Met. Soc., 93, 494-500

Measurements Relating to the Heterogeneous SO_2-Oxidation in the Atmosphere

W. Jaeschke

Referat für Umweltschutz, J. W. Goethe-Universität Frankfurt/Main, FRG

S u m m a r y.

Measurements relating to the heterogeneous SO_2- oxidation in the atmosphere, have been performed in the field and in the laboratory. Besides phenomenological measurements of the chemical composition of heterogeneous systems, field experiments have been performed to get a previous approximate calulation of the magnitude of a first order removal coefficient of SO_2 in the real atmosphere. Based on the results of the field experiments a reaction chamber was designed for laboratory measurements to study the removal coefficient as a function depending upon the chemical properties of selected heterogeneous systems.

I n t r o d u c t i o n.

Cloud droplets and aerosol particles play an important role in the heterogeneous conversion of atmospheric SO_2 to sulfuric acid and sulfate. The rate determining influences for these processes may be either of physical or of chemical character. While theoretical meteorologists (F.Herbert) are concerned with the microphysics of clouds, our group studies the chemical aspects of the SO_2 removal processes in the aqueous phase of hydrometeors or on the surface of aerosol particles. It is generally agreed that the SO_2 removal rate is proportional to the SO_2 concentration present in the aqueous phase. For purposes of comparison a pseudo - first order removal coefficient K_0 has been determinated to describe the chemical kinetics of the SO_2-oxidation in the atmospheric aqueous phase (Hegg and Hobbs 1978).

$$- \frac{dSO_2}{dt} = K_0 (SO_2) \qquad (1)$$

The scientific problem is now to investigate the influence of chemical properties of various heterogeneous systems on the SO_2 removal coefficient. Measurements concerning this problem were carried out under three main aspects.

1)
Field experiments enabled us to get a previous approximate calculation of the magnitude of the overall removal rate of SO_2 in the real atmosphere.
2)
Phenomenological measurements were performed in order to gather more

more informations about the concentration and distribution of trace substances in cloud and precipitation water.

3)
A reaction chamber was designed in which airborne heterogeneous systems can be generated in the presence of SO_2 mixing ratios in the lower ppb range.

Field Experiments.

In order to specify the magnitude of the removal coefficient K_0 in the real atmosphere field experiments were performed in the plume of, Mt. Etna in Southern Italy.

The measurements of trace substances in the vicinity of volcanoes were possible only because special sampling and analyzing methods had been developed. The chemistry of these methods is schematically shown in Table 1 (A and B). The discontinuously working instruments which have been described extensively in an earlier paper (Meixner and Jaeschke, 1980) could be used in small aircrafts with limited energy capacities, as well as under such extraordinary field conditions like in the vicinity of the volcanoes Etna, Stromboli and Vulcano.

Table 1 Detection methods applied in the vicinity of volcanoes

A) DISCONTINUOUS

COMPOUND	ACCUMULATION	ANALYSIS
H_2S	$H_2S + AgNO_3 \rightarrow Ag_2S + HNO_3$	FLUORESZENZ QUENCHING
SO_2	$2SO_2 + Na[HgCl_4] \rightarrow Na[Hg(SO_3)_2] + 4HCl$	CHEMI-LUMINESZENZ
SO_4^{--}	PARTICLE FILTER	ISOTOPIC DILLUTION

B) CONTINUOUS

COMPOUND	TYPE OF DETECTOR	REMARKS
CO_2	INFRARED ABSORPTION	BOTH PRESSURE DEPENDEND
SO_2	FLAME PHOTOMETRIC	

Following a preliminary survey extensive field investigations were carried out in September 1978 and June 1979. During these investigations the concentrations of the sulfur compounds H_2S, SO_2 and SO_4^{2-}, as well as the CO_2- concentration were measured in the plume of Mt. Etna. A summary of the concentration data, determined near the Bocca Nuova crater rim, considered one of the central craters of Mt. Etna is compiled in Table 2. The concentrations of the heavy metals and of the water vapor in the plume were taken from measurements by Buart - Menard and Arnold (1978) and Tazieff (1970). Our results demonstrate that the plume of the volcano is an ideal system for the inves-tigation of the hete-rogeneous SO_2- oxida-tion in the presence of heavy metals and water droplets.

Table 2 Concentration of Sulfur compounds, heavy metals, and water vapur determined near the Bocca Nuova Crater rim.

Comp.	Sept. 1978 [μg/m³]	June 1979 [μg/m³]	Metal	[μg/m³]
H_2S	0,54 ±0,03	330 ± 210	Fe	10
SO_2	$(22,84 \pm 2,3) \cdot 10^3$	$(57,10 \pm 5,7) \cdot 10^3$	Cu	3,2
SO_4^{--}	$6,07 \cdot 10^3$	$4,05 \cdot 10^3$	Mn	1,3
CO_2	[ppm (V)]	[ppm (V)] 1077±91	Pb	1,2
H_2O	[Vol.%] 3-40		Cd	0,09

The investigations were carried out during aircraft ascents. Fig. 1 represents a schematic design of the flight pattern tracked in the measuring region. While crossing the Etna plume at various altitudes and distances from the crater, the concentration distributions of the individual trace substances were measured three-dimensionally: longitudinally along the x- axis, crosswise to the wind direction along the y-axis, and vertically following the z-direction. The aircraft departed from Catania airport and returned to the same airport after each flight.

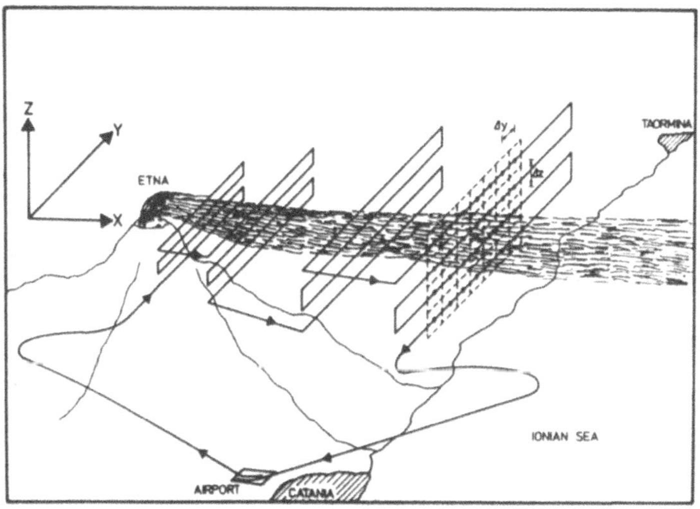

Fig. 1 Flight pattern tracked in the measuring region at Mt. Etna

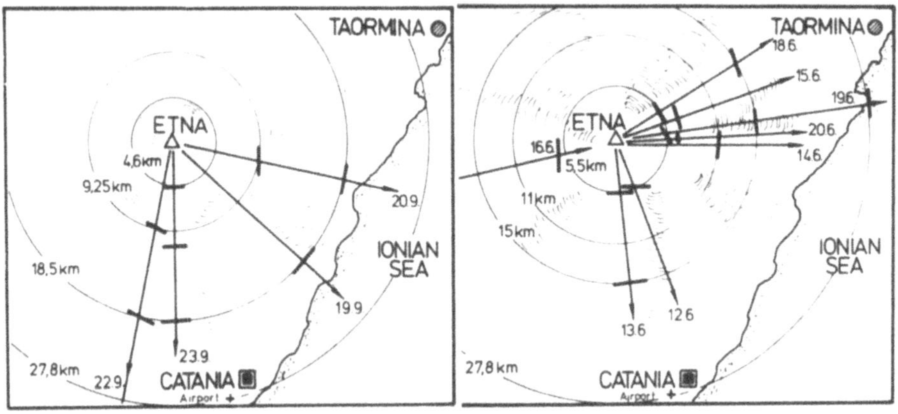

Fig.2 Flightprogramme a) Sept. 1978 b) June 1979

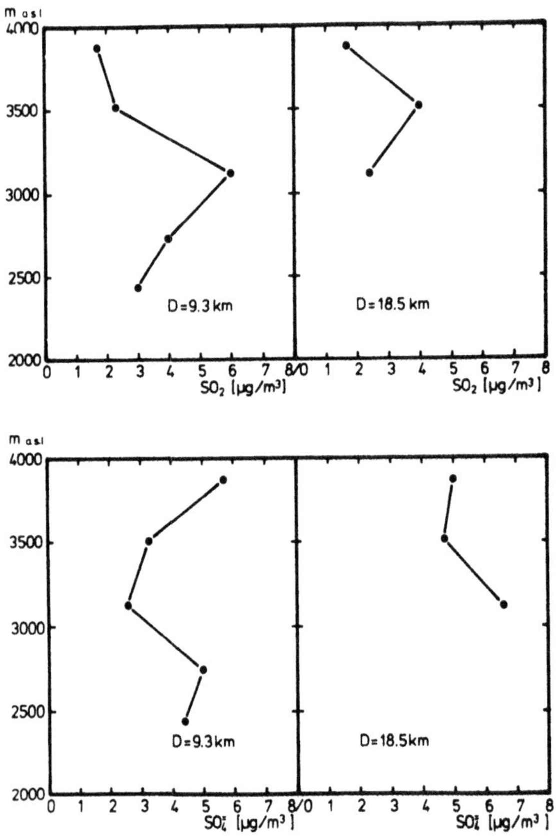

Fig. 3 Vertical profiles of the SO_2- and Sulfate- concentration

A summary of all measuring flights carried out in Sept. 1978 and June 1979 is schematically shown in Fig. 2a and 2b. The mean wind directions during the flights are represented by arrows, while the traverses through the plume are represented by short, thick cross- beams. The various distances from the point source in the central crater region are given by using circle coordinates. In Fig. 2b additionally, a measuring flight is indicated during which background concentrations were measured on June 16, 1979, while passing the leeside of Mt. Etna.

Some typical examples for the concentration profiles of SO_2 and sulfate obtained in three dimensions during flights throughout the plume of Mt. Etna, are shown in Fig. 3 - 5. First of all in Fig. 3, vertical profiles of the

SO_2 and sulfate con-centrations are given which had been measured at various distances from the crater. It is noteworthy that in both instances the distributions show a contrasting behavior. Maxima of the SO_2- concentration occur at the axis of the plume, and at the same locations the sulfate concentration reaches its minimum, while the reverse is true at the plume margins. This shows that the processes of sulfate formation during the spreading of the plume are more intense at the margins than in its center. The concentrations given in Fig. 3 are mean values, which were obtained during the traverse flights through the plume by using discontinuous as well as continuous (i.e. for SO_2) measuring methods.

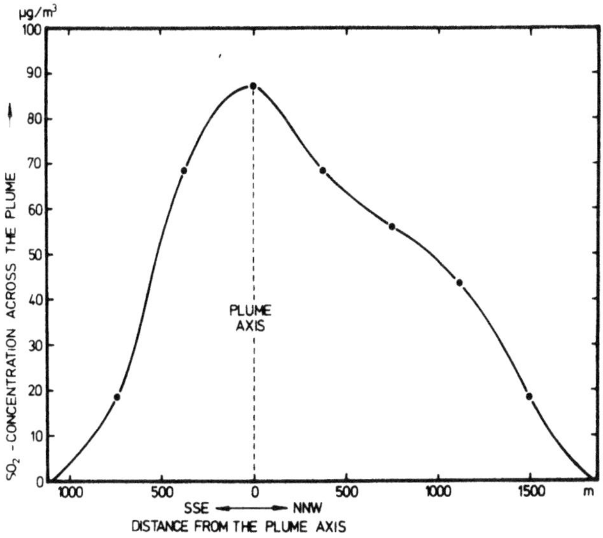

Fig. 4 Horizontal profil of the SO_2- concentration

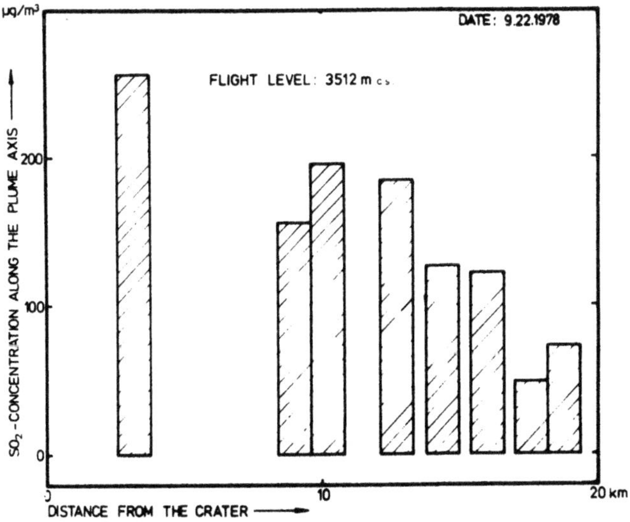

Fig. 5 SO_2- concentration data as the were obtained along the plume axis.

Fig. 4 represents a typical example for the continuous registration of the SO_2- concentration profile during a crosswind flight. In Fig. 5 the SO_2- concentration values are reported as they were obtained on a flight along the plume axis. Obviously the concentration does not at all decrease continuously with increasing distance. Inconsistancies do occur. This is due to a vertical meandering of the plume which was observable with the naked eye and which produced certain difficulties during the execution of the flight patterns.
In order to determine the removal rate coefficient of the individual components the mass transport in the plume of the volcano had to be calculated from the measured atmospheric concentration distributions. Assuming that the whole massflux is registered during a measuring flight in a given transverse cross section of the plume one can express the decrease of the massflux between transporttime t_1 and t_2 as an exponential function, which is only determined by the chemical removal coeffcient k_o.

$$\phi_{t_2} = \phi_{t_1} \, e^{-K_o \Delta t} \qquad (2)$$

In order to calculate the massflux as a first step two- dimensional concentration fields were drawn by combining the horizontal and vertical concentration profiles measured in various distances. The position of such a field is shown in Fig. 1 as a transverse cut area in the vertical y,z-plane. The grid divides this section into small area elements $\Delta y \Delta z$, which are associated with average concentrations $C_i(y,z)$. Taking Fig. 4 as an example for a horizontal profile of the SO_2- concentration in the plume, the dots illustrate how these average concentrations were determined graphically from the mea-suring data for equidistant sections of the corresponding crosswind flight.

By linear interpolation between the individual C_i- values the isolines of the concentration distribution in a given transverse could be drawn. Fig. 6a,b and 7a,b, as examples of such a graphic evaluation of the measurements, represent an image of the shape of the plume. In Fig. 6a,b a clear inhomogeneity is noticeable, the plume tending to bifurcate with increasing distance to the source. Due to the weak fuming activity of Mt. Etna in Sept. 1978 the plume attained a horizontal diameter of only 3 km at a

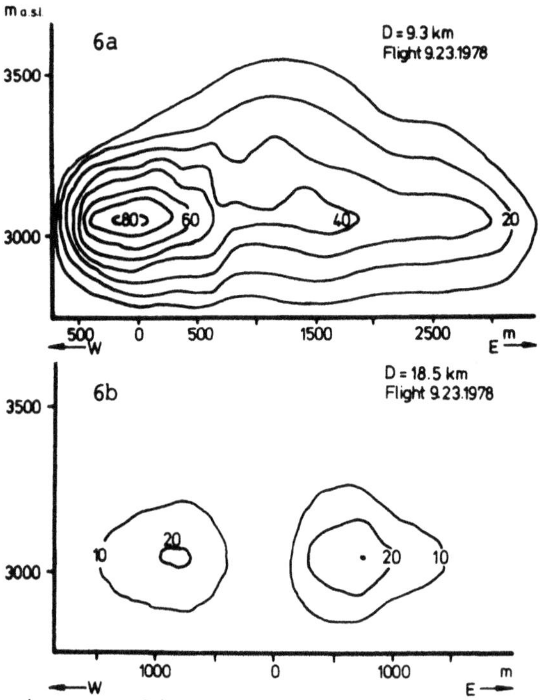

Fig. 6 Isolines of the SO_2- concentrations in the plume of Mt. Etna (ppb)

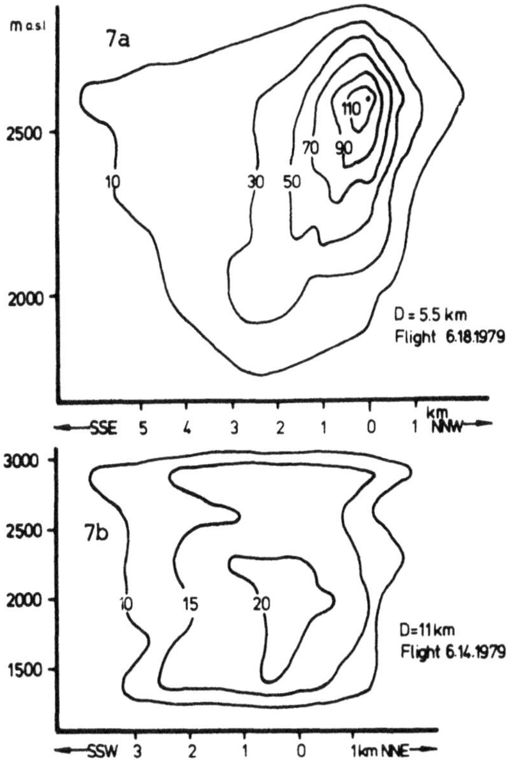

Fig. 7 Isolines of SO_2- concentration in the plume of Mt. Etna (ppb)

distance of 9.3 km with an maximum SO_2- concentration of 84 ppb (240 ug/m^3 STP) in its center. The increased fuming activity of Mt. Etna in June 1979 is also reflected in the shape of the plume. Besides the higher concentration for SO_2 in the center of the plume, a larger horizontal diameter of approximately 6 km was attained, as well as a more compact plume shape. These examples demonstrate representative concentration conditions in the plume of the volcano which showed only slight variations during the corresponding measuring periods because of the continuous character of the fuming activity. Devious behavior was registered merely on June 19 and 20, 1979, after a part of the crater rim of Bocca Nuova had broken off.

Having checked the plausibility of the gathered $C_i(y,z)$ data with the help of the described graphic method, eq. (3) was used to calculate for each sulfur compound the individual massflux ϕ which advanced through a transverse cross section at a given distance D to the crater:

$$\phi(D) = \sum_{i=1}^{N} C_i(y,z)\, \Delta y \Delta z \quad \bar{u} \quad \cos\Theta \tag{3}$$

with
N = total number of area elements $\Delta y \Delta z$
$\bar{u}$ = mean wind velocity at the flight level z
Θ = mean angle between area normal and mean wind direction

The number N of area elements varied between 20 and 40, depending upon the quantity of the data available. The values of the mean wind velocity u and the wind direction were known as they had been determined locally with pilotballoon-ascents and with radio-sondes at the nearest aerological stations Qrendi (Malta), Trapani and Brindisi. Table 3 contains the calculated massfluxes of the individual sulfur compounds H_2S, SO_2 and sulfate together with the corresponding travel time t, which results applying eq. (4):

$$t = D/u \tag{4}$$

Table 3 Massfluxes and Transport time

DATE	Ø(t/d)			t(min)
	$Ø_{H_2S}$	$Ø_{SO_2}$	$Ø_{SO_4^{--}}$	
1978				
9/20	2.0		190.0	19.4
		43.6		20.8
		14.6		38.0
	1.8		154.0	40.6
9/22	1.3			13.7
			198.0	14.8
	0.9			27.3
			166.0	27.9
	1.8			41.0
9/23			87.6	11.9
	0.8			15.8
			70.7	27.9
		65.2		28.2
	1.0			30.0
		28.7		33.9
			68.4	56.4
		5.9		58.5
		15.0		58.5
1979				
6/12		433.3	314.8	10.7
6/13		137.1	61.5	31.0
6/14	5.1	581.7	127.4	9.8
6/15		596.0	120.0	7.1
6/16				
6/18	3.6	574.7	92.6	6.7
	1.9	522.5	35.4	14.2
6/19		1550.0	153.6	2.6
		1300.0	77.1	13.4
6/20		840.0	15.1	7.9

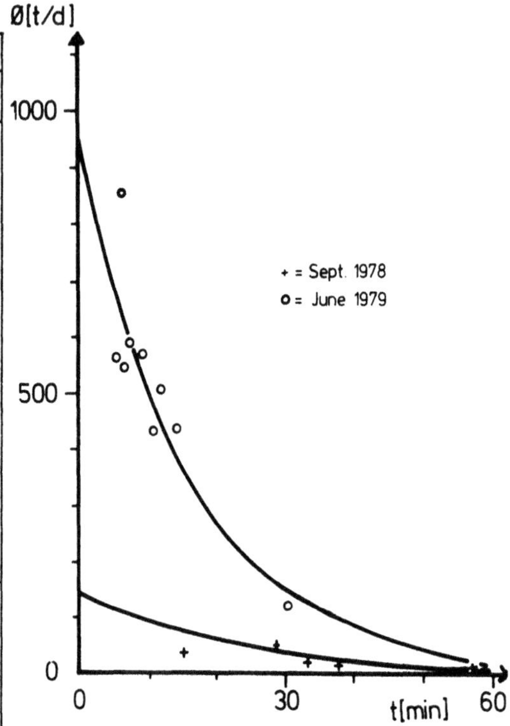

Fig. 8 SO_2- massfluxes as a function of transport time +,o measured values

The calculated massfluxes are now used to determine the removal coefficient by equation (2) with the help of a regression calculation. In Fig. 8 the calculated functions $Ø=f(t)$ are given together with the measured mass fluxes. The obtained removal coefficients are compiled in Table 4. Obviously they are both in the same order of magnitude of 10^{-4} sec^{-1}.

Table 4 Calculated removal coefficients

	1978	1979
Method	$K_o(sec^{-1})$	$K_o(sec^{-1})$
Massflux	$7.8x10^{-4}$	$10.0x10^{-4}$
Tracer	$6.1x10^{-4}$	-

In order to examine the calculated values, the removal coefficient was also determined from the horizontal concentration distribution. Assuming an exponential decrease of the SO_2- concentration eq. (5) can be applied since the two measuring points on the plume axis correspond to the travel times t_1 and t_2.

$$(SO_2)_{t_2} = (SO_2)_{t_1}\, e^{-Kt} \qquad (5)$$

The overall dilution rate constant K implicates diffusional, chemical and physical processes (the latter being mainly dry and wet deposition) which effect dilution or removal of the substance during its transport in the plume. In a first approximation and for small values of .t, K can be expressed simply by the sum of the diffusion rate constant K_D and the removal coefficient K_O

$$K = K_D + K_O \qquad (6)$$

In order to determine the dilution of the transported substances in the plume by diffusion processes, CO_2 measurements were carried out simultaneously (see Table 1). Since with increasing travel time, the concentration decline of this inert gas is governed by diffusion only, one can write eq. (7), using eq. (5) and (6).

$$(CO_2)_{t_2} = (CO_2)_{t_1}\, e^{(-K_D \cdot t)} \qquad (7)$$

If the physical removal process concerning the reactive substance (in this case SO_2) is negligible, i.e. if the following conditions are fulfilled:

a) no dry deposition at the slope of the volcano,
b) no rain- or washout processes, i.e. no penetration of the plume through clouds between the volcano and the aircraft,

as a consequence the removal coefficient K_O can be determined by combining eq. (5), (6) and (7) as follows:

$$\frac{((H_2S)/(CO_2))_{t_2}}{((H_2S)/(CO_2))_{t_1}} = e^{(-K_O \cdot t)} \qquad (8)$$

The coefficient K_O for a pseudo - first order reaction of SO_2 in the plume of Mt. Etna was calculated accordingly for Sept. 1978, as for that time the conditions a) and b) were fulfilled. The value of $6.1 \cdot 10^{-4}$ sec^{-1} is also shown in Table 4. It is in good agreement with the values obtained by the massflow method.

Phenomenological Measurements.

It is obvious that the conditions of the above field experiments are not representative for the atmosphere. The same is true for the calculated removal coefficients. In order to study the influence of atmospheric conditions on the removal coefficient, it is necessary to investigate the chemical composition of various heterogeneous systems. Therefore phenomenological measurements of the concentration and distribution of the trace substances in hydrometeors and aerosol particles were performed in the Rhein-Main-Area and above the Atlantic Ocean.

An example for this activity is shown in Fig. 9. The concentrations of heavy metals and sulfur compounds were measured in clouds during aircraft ascents above the Rhein- Main-Area and in the precipitation at the ground simultaneously (Möller 1979). The measured concentrations of the individual components are plotted in a block diagram. The mean values in precipitation are slightly higher than those measured in the clouds. This is in agreement with the theories of rainout and washout processes. Furthermore the absolute values show, that the atmospheric concentrations of sulfur compounds and heavy metals in this industrialized region are much lower than those, which so far have been simulated in laboratory experiments (e.g. Runca Köberich 1980).

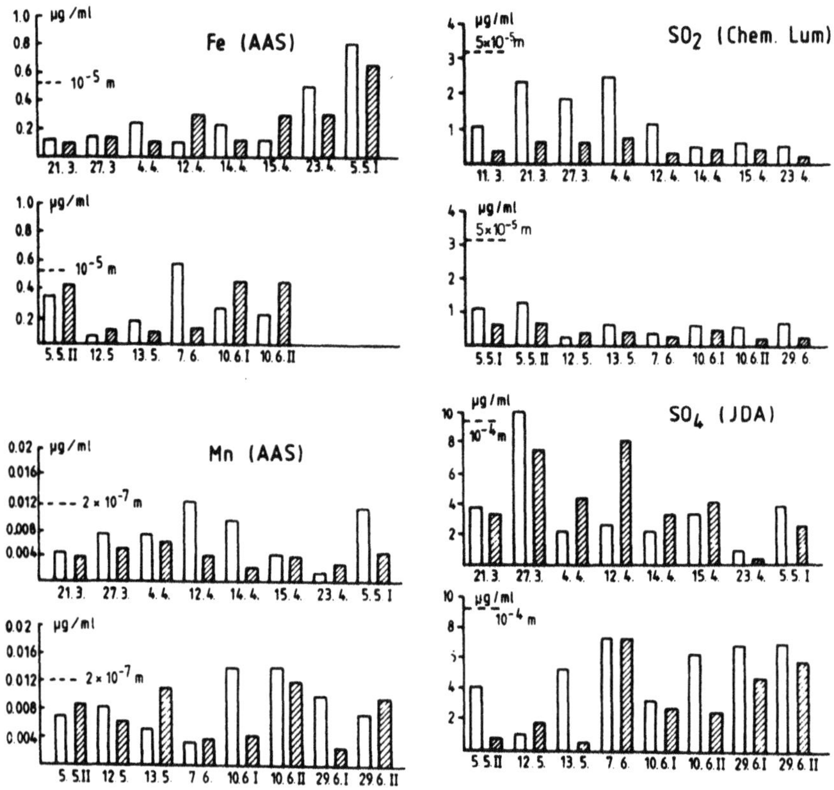

Fig. 9 Simultaneously measured concentrations of heavy metals and sulfur compounds in clouds ▨ and precipitattion ☐ at the ground.

L a b o r a t o r y   M e a s u r e m e n t s

Based on the experiences of the field experiments and of phenomenological measurements a reaction chamber was designed. With this device it should be possible to simulate heterogeneous systems with realistic concentrations and mixtures of trace substances.

These experiments should enable us to gain valves of a first order removal coefficient in the range of $10^{-4} sec^{-1}$.

Fig. 10 shows a scheme of the reaction chamber and the whole experimental set-up, which is used to simulate selected atmospheric conditions. With the aid of a permeation tube and a microbalance a steady state SO_2- concentration of e.g. 30 $ug/m^3 \pm 0{,}1\ ug/m^3$ is generated in the chamber by a flow system (Beltz 1980). Upon reaching stabilized concentration conditions an aerosol of known chemical composition and known size distribution is blown into the chamber. Variable transport times are set by adjusting the flow rate. At the end of the chamber the SO_2 removal is registered with a chemiluminescenc e method. With this method a decrease of the SO_2- concentration is detectable down to 0.1 ug/m^3 (Meixner and Jaeschke 1981). Thus considering a flow rate of 1.5 m^3/h, which corresponds to a transport time in the chamber of 170 sec, a removal rate coefficient in the order of 10^{-4} sec^{-1} can be determined.

This chamber in conjunction with field experiments is suited to investigate the influence of the composition of various heterogeneous systems on the SO_2 removal coefficient. We expext that this approach will lead to the determination of real rate constants of higher orders.

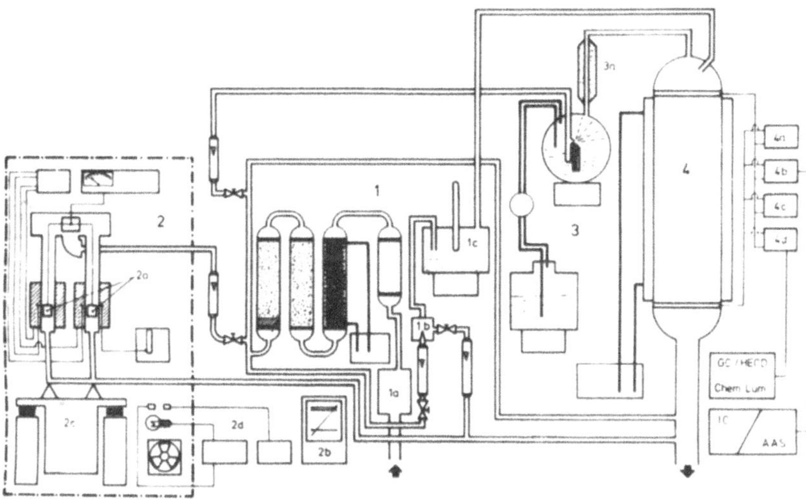

Fig. 10 Simulation Device, Schematic Diagramm

1 Clean-Air and Gas-Dilution Device with Filtercolumns
1a Hight-Volume-Pump
1b Mixing-Head for Prepapation of known SO_2-Concentrations
1c Water-Vapur-Source
2 Microbalances with Temperature-Controlled Weighing Tubes
2a Permeation-Tube with Conter-Weight
2b Four-Channel-Recorder
2c Table for Microbalances
2d Thermostat
3 Aerosolgenerator with Atomizer and Supply- Container
3a Diffusiondryer
4 Temperature Controlled Reactionchamber
4a Temperatur and Humiditymeter
4b Impactor with Analysing Procedur by Ion- Chromatography and AAS
4c Particle Counter
4d SO_2-Analyses by Chemiluminescence or GC/HECD

L i t e r a t u r.

1 Beltz, N. 1980
Eine dynamische gravimetrische Methode zur Eichung luftchemischer Analysegeräte
Dipl. Arbeit, Inst. f. Meteorologie und Geophysik, Uni. Frankfurt

2 Buat-Menard, P. and Arnold, M. 1978
The heavy metal chemistry of atmospheric particulate matter emitted by Mount Etna vulcano
Geophys. Res. Lett. 5, 245-248

3 Hegg, D.A. and Hobbs, P.V. 1978
Oxidation of Sulfur dioxide in aqueous systems with particular reference to the atmosphere.
Atm. Env. 12, 241-255

4 Meixner, F.X. and Jaeschke, W. 1981
The detection of low atmospheric SO_2- concentration with a Chemiluminescence Technique
Int. Jour. of Env. Anal. Chemistry 10, 51-67

5 Möller, H. 1979
Experimentelle Untersuchung zur Auswaschung von Spurenstoffen aus der Atmosphäre durch natürlichen Niederschlag.
Dipl. Arbeit Inst. f. Meteor. und Geophysik, Universität Frankfurt

6 Runca-Köberich, D. and Georgii, H.W. 1980
Removal and Transformation processes for sulfur dioxide in the atmosphere
Developments in atmos. science 11
Edited by A. Longhetto, 273-285

7 Tazieff, H. 1970
New investigations on eruptive gases
Bull Volc. 34, 421-438

Contribution of Tropospheric SO_2 to the Stratospheric Aerosol Layer

F. X. Meixner[1)]

Institut für Meteorologie und Geophysik, J. W. Goethe-Universität Frankfurt/Main, FRG

A b s t r a c t. Measurements of SO_2 mixing ratio in the upper troposphere and lower stratosphere have been performed with a sensitve chemiluminescence method which permits the detection of atmospheric SO_2 traces below 10 pptv. Results of 9 aircraft ascents over Europe from 1978 - 1980 are reported. It is shown that the meteorological conditions at the tropopause level have an important influence on the observed SO_2 mixing ratio. The weak vertical SO_2 gradients in the upper troposphere suggest only a small flux of tropospheric SO_2 into the stratosphere. Furthermore, increasing SO_2 mixing ratios within the first kilometers of the stratosphere give strong support to a stratospheric SO_2 source due to the oxidation of organic sulfur compounds (COS, CS_2). SO_2 measurements are discussed in the light of improved 1 D models concerning the stratospheric sulfur budget. The results clearly show that tropospheric SO_2 is of only minor importance for the non-volcanic formation of the stratospheric aerosol.

1. INTRODUCTION

Junge et al. (1961) first suggested that tropospheric sulfur bearing trace gases, above all the anthropogenic SO_2, are the precursors of the global stratospheric sulfate layer at times of minor volcanic activity. This remained the accepted view until 1976. However, renewed discussion about the formation and persistence of the stratospheric aerosol was stimulated by the papers of Hofmann et al. (1976) and Crutzen (1976). After a review of their in-situ measurements of stratospheric particles and, on the basis of both sulfur budget considerations and 1 D model calculations, Hofmann et al. doubted that a non-volcanic, quasi-steady state tropospheric SO_2 source would be the only reason for the observed global stratospheric aerosol layer. Concurrently Crutzen proposed that the photodissociation of carbonyl sulfide (COS) may provide an important stratospheric source of SO_2 and that further conversion of this SO_2 to H_2SO_4 may contribute significantly to the non-volcanic background sulfate layer.

However all considerations and model calculations before 1976 lacked measured stratospheric SO_2 data. This was due to the lack of sensitivity of the SO_2 available methods to detect the extremely low SO_2 mixing

[1] Present affiliation: Kernforschungsanlage Jülich GmbH, Institut für Chemie 3: Atmosphärische Chemie, Postfach 1913, D-5170 Jülich, F.R.G.

ratios expected above the tropopause. After the development of a very sensitive chemiluminescence method (Stauff and Jaeschke, 1975) the first SO_2 measurements from the lower stratosphere became available in 1976 (Jaeschke et al. 1976).

The SO_2 measurements reported in this paper were performed using this method which consists of separate sampling and analysis stages applying a wet chemical filter procedure. The method was improved and modified (Meixner, 1981) for the application in commercial high flying aircraft (Hawker Siddely 125, Learjet 24 D). The detection limit of the method is (8 ± 2) pptv SO_2 (3σ).

2. AIRCRAFT ASCENTS

A total of 9 flights (see Fig. 1) were performed over western Europe (40^0 - 60^0 N, 10^0 W - 12^0 E). Since the available aircraft could reach the lower stratosphere at 40^0 - 60^0 N only during the cold season, it was decided to perform the ascents mainly during winter. Planning the flights, one must take the following point into account: a) at a given location a complete exchange of air masses throughout the whole troposphere and lower stratosphere can occur within a period of several hours to a day, b) cyclogenetically active (polar-front) jet streams accompanied by a so called "folded tropopause" frequently appear and can lead to pronounced stratospheric-tropospheric exchange processes (see Reiter, 1975). Therefore, in order to assign the measurements to uniform air masses the duration of the ascents was limited to a period of 3 hours. Furthermore, additional measurements of the ozone mixing ratio, air temperature and pressure combined with the evaluation of the upper air records of the European radiosonde-stations were used to identify the stratospheric or tropospheric origin of the obtained SO_2 mixing ratios.

The measurements are classified with respect to the behaviour of the tropopause. Principles of classification are discussed in more detail elsewhere (Meixner, 1981).

The results of the ascents from 1978 - 1980 are shown in Fig. 1. Measurements on 26 May and 9 Dec 1980 are not discussed in this paper because they were made after the eruptions of the volcano Mount St. Helens which released substantial amounts of SO_2 in the stratosphere (see Meixner, 1981; Meixner et al., 1981).

3. RESULTS

3.1 MEASUREMENTS IN THE UPPER TROPOSPHERE

Between the 500 mb level and the actual tropopause the SO_2 mixing ratio was found to be < 100 pptv in all profiles. According to the different conditions encountered during the ascents with respect to seasons, location and the meteorological behaviour of the upper troposphere, SO_2 mixing ratios range between 19 - 85 pptv. Considering only the ascents on 25 Apr 1978, 20 Sep 1979 and 24 May 1980, which were performed below an undisturbed tropopause (i.e. respective air masses could clearly be identified as tropospheric) the vertical gradient of SO_2 mixing ratio depends strongly on the distance to the tropopause (see Tab. 1). The gradient at the tropopause is small. Since the vertical flux is propor-

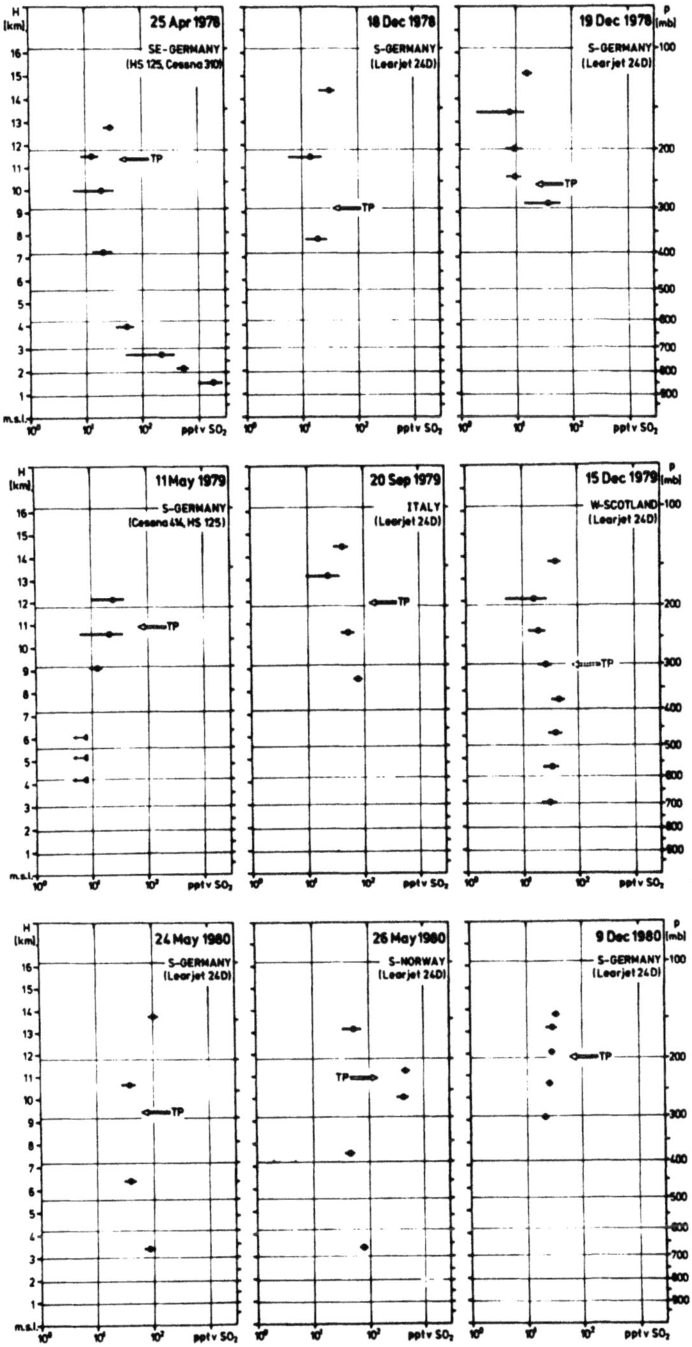

Fig. 1: *Vertical profiles of SO_2 mixing ratio over Europe, 1978 - 1980. Error bars correspond to 1σ. The tropopause is indicated by the arrow.*

Tab. 1: *Vertical SO_2 gradients in the upper troposphere at conditions of an undisturbed tropopause (average values of the ascents of 25 Apr 1978, 20 Sep 1979, 24 May 1980)*

average vertical SO_2 gradient	average distance to the tropopause
pptv/km	km
-22	-5
- 9	-3
- 0.5	-1

tional to the gradient one might expect only a very small amount of SO_2 flux from the troposphere to the stratosphere. Obviously the estimate of this flux requires the knowledge of the effective eddy diffusion coefficient at the tropopause level (see below).

3.2 MEASUREMENTS UNDER CONDITIONS OF A DISTURBED TROPOPAUSE

As previous observations have indicated (Georgii and Meixner, 1980) the SO_2 profiles obtained in the effective range of the polar-front jet stream are characterized by repeated and well-marked changes of the sign of the vertical gradient. Since there are no sources of SO_2 in the upper troposphere which could cause positive gradients in the tropopause region, the measured structures of the SO_2 profiles can only be explained by atmospheric transport most likely of stratospheric air into the troposphere. Earlier measurements of ^{90}Sr, ^{95}Zr and ozone within the region of cyclogenetically active jet streams showed similarly structured layers which are unambiguously identified as resulting from stratospheric-tropospheric exchange processes (e.g. Danielsen, 1968, Danielsen et al. al. 1970). The mechanisms of this exchange can be described as follows (Danielsen, 1975). Stratospheric air is transported from the cyclonic (northern) side to the anticyclonic (southern, tropospheric) side of the jet stream along downward sloped layers of high stability. The corresponding upward transport of tropospheric air generally takes place on the anticyclonic side of the jet stream.

As an example, the results of the ascent performed on 15 Dec 1979 west of the Scottish coast will be discussed in more detail. The results (see Fig. 2 show increasing SO_2 mixing ratios between 3 km - 7.6 km (flight levels 8 - 5), which are followed by decreasing values (9.1 km - 12.1 km, levels 4 - 2), while at the highest level (#1, 13.8 km) the mixing ratio is raised again. A clear indication of stratospheric-tropospheric transport could not be deduced from the actual (15 Dec 1979) charts of tropopause height or from the detailed evaluations of upper air records of the respective radiosonde-stations. But it is evident from the simultaneously measured O_3 mixing ratio that an intrusion of stratospheric air into the middle troposphere had occurred. The complexity of the spatial O_3 distribution is shown in Fig. 3. Ozone mixing ratios which are typical for the lower stratosphere (0.7 - 1.2 ppmv) are observed at flight levels 1 - 3 (13.8, 12.1, and 10.7 km), while typically lower tropospheric values (< 0.1 ppmv) are found at level 6 - 8 (6.1, 4.6, and 3.0 km). However, measurements show a sudden increase of O_3 mixing ratio in flight level 4 from about 0.3 ppmv to 3 ppmv and a subsequent drop from 3 ppmv to 0.15 ppmv in level 5. Such limited layers of thick-

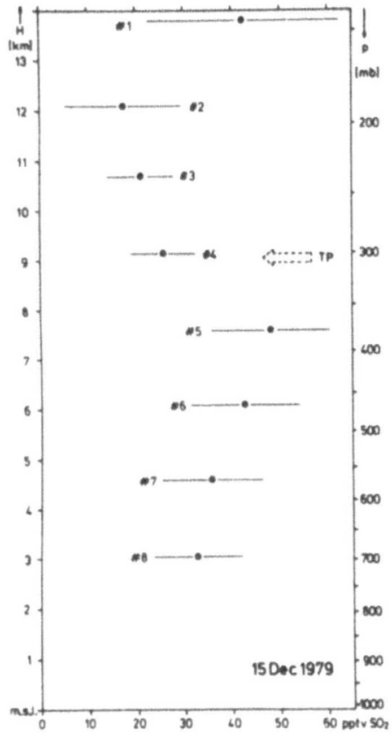

Fig. 2: *Vertical SO_2 distribution on 15 Dec 1979. Aircraft ascent west of the Scottish coast.*

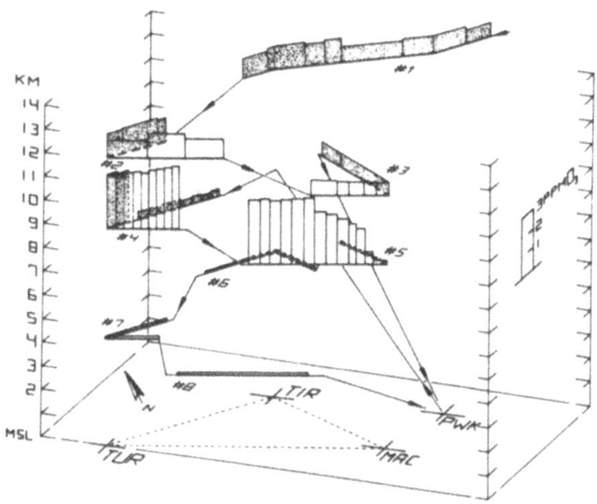

Fig. 3: *O_3 distribution in 3 dimensions along the flight track. PWK: Prestwick (55.5° N, 4.6° W), MAC: Machrihanish (55.4° N, 5.7° W), TIR: Tiree (56.5° N, 6.9° W), TUR: turning point (57° N, 10° W).*

ness of unusually high O_3 mixing ratio at a height of approx. 8 km clearly characterize an intrusion of stratospheric air. This is furthermore confirmed by an estimation of isentropic trajectories (Meixner, 1981), by which air parcles could be spatially traced back for a period of 24 h to the middle of the North Atlantic in all flight levels. Upper air analyses of the German Weather Service showed that a jet stream system having three branches at different altitudes was located in that area. The vertical distribution of potential temperature and relative humidity as a function of time (Meixner, 1981) at the location of weather ship L (57° N, 20° W) were used to identify the organized transport of stratospheric air into the troposphere. The diminished SO_2 mixing ratios between 9.1 - 12.1 km should be assigned to air masses originating from the anticyclonic side of the jet streams where massive upward transport of tropospheric air occur (Shapiro et al., 1980). The isentropic trajectory for the highest flight level (13.8 km) was always located deep in the stratosphere and not affected by the afore mentioned exchange processes.

3.3 MEASUREMENTS IN THE STRATOSPHERE

Because of the ceiling of the available aircraft of about 14 km, measurements in the (undisturbed) lower stratosphere are somewhat restricted by the variable and seasonally dependent height of the tropopause (at 40° - 60° N: 9 - 14 km).

However, all ascents performed over an undisturbed tropopause show one common result: the vertical gradient of the SO_2 mixing ratio between the highest two levels is always positive (see Tab. 2). Using cryogenic trapping and gaschromatographic (FPD) analysis for the detection of atmospheric sulfur compounds over California and Alaska, Inn and Vedder (1981) also observed a lack of decrease in SO_2 mixing ratio even at altitudes up to 20 km (~ 50 pptv). The results in Tab. 2 supplemented by the measurements of Inn and Vedder, are in striking contrast to the earlier model calculations of the stratospheric sulfur budget (e.g. Junge 1974, Harker, 1975). Assuming that tropospheric SO_2 is the only gaseous precursor of the stratospheric aerosol, it was expected from simple budget consideration that the SO_2 mixing ratio had to significantly decrease within the first few kilometers above the tropopause. The observation of increasing SO_2 mixing ratios in the lower stratosphere implies a stratospheric SO_2 source.

Tab. 2: *Vertical SO_2 gradients in the stratosphere at conditions of an undisturbed tropopause*

date	vertical SO_2 gradient pptv/km	distance to tropopause km
25 Apr 1978	+12	0.73
18 Dec 1978	+ 6	3.59
19 Dec 1978	+ 5	3.98
20 Sep 1979	+15	1.83
24 May 1980	+24	2.54

4. DISCUSSION

The results reported in section 3 will be discussed in the light of two improved 1 D models given by Turco et al. (1980) and Crutzen (1981) considering the vertical distribution of atmospheric sulfur compounds. A schematic description of the chemical, physical and meteorological processes that are included in the models that govern the sulfur budget in the region of the tropopause is shown in Fig. 4. The main reactions leading to the production and destruction of SO_2 are summarized in Tab. 3. The photodissociation of CS_2 and COS seems to be of minor impor-

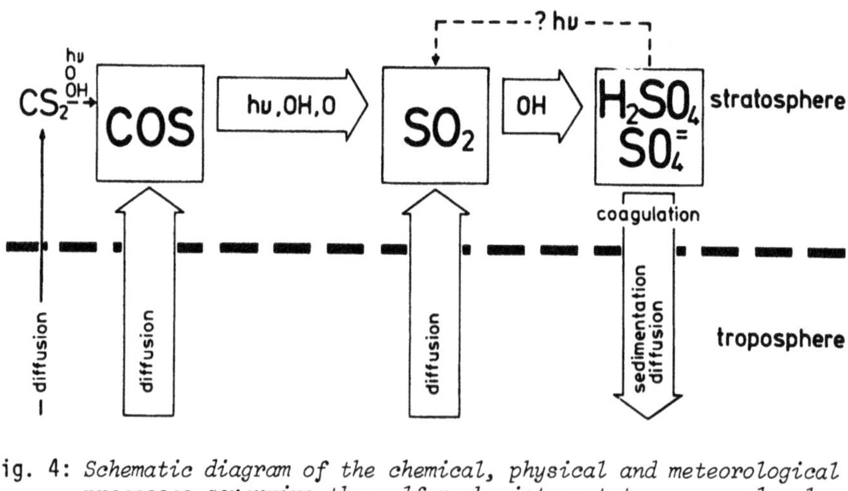

Fig. 4: *Schematic diagram of the chemical, physical and meteorological processes governing the sulfur chemistry at tropopause level*

Tab. 3: *Reactions and kinetic data for gaseous sulfur compounds at the tropopause level*

reaction		reaction rate cm^3s^{-1}	reference
$COS + OH$	$\rightarrow SH + CO$	5.7×10^{-14}	KURYLO (1978)
		8.8×10^{-15}	RAVINSHANKRA et al. (1980)
$CS_2 + OH$	$\rightarrow SH + COS$	1.9×10^{-13}	KURYLO (1978)
		0.3×10^{-14}	IYER and ROWLAND (1980)
$COS + O$	$\rightarrow SO + CO$	$2.6 \times 10^{-11} \exp(-2250/T)$	BAULCH et al. (1980)
$CS_2 + O$	$\rightarrow CS + SO$	$5.8 \times 10^{-11} \exp(-700/T)$	BAULCH et al. (1980)
$SO_2 + OH + M$	$\rightarrow HSO_3 + M$	$\frac{(8.2 \pm 2.27) \times 10^{-13} \times [M]}{(0.79 \mp 0.34) \times 10^{-18} + [M]}$	MOORTGART and JUNGE (1977)
		$\frac{2 \times 10^{-12} \times A}{1 + A} \times 0.6^{(1+(\lg A)^2)^{-1}}$; $A = 1.5 \times 10^{-19} (\frac{300}{T})^{2\ 9} \times M$	NASA (1981)

tance in the region of the tropopause (Crutzen, 1981, private communication) and is therefore omitted. The most importance oxidant in the sulfur chemistry is the OH radical and the second critical parameter is undoubtedly the vertical distribution of the effective eddy diffusion coefficient (K_Z) in the region of the tropopause. The OH- and K_Z-profiles are different in both models. A comparative discussion is given by Meixner (1981). SO_2 data obtained under undisturbed conditions of the tropopause and supplemented by the results of Inn and Vedder (1981) are shown in Fig. 5 together with the model calculations. Keeping in mind that there is a basic problem in comparing (average) model results with (individual) measured values, it can be recognized that a) measured values, especially during winter season, are occasionally lower than the calculated data by more than a factor of 2, b) the pronounced positive gradients of SO_2 mixing ratio observed in the lower stratosphere have no equivalent in the calculated profiles, and c) model results do not fit the data of Inn and Vedder at the 20 km level. These discrepancies are considered at length by Turco et al. (1980) and seem to be a consequence of the proper choice of the OH- as well as the K_Z-profile. Nevertheless, for the purpose of this paper it should be emphasized that the observed increasing SO_2 mixing ratio in the lower stratosphere may be only explained, qualitatively, with the introduction of a stratospheric SO_2 source most likely the oxidation of COS and CS_2. Both compounds originate from biological, volcanic and industrial sources (e.g. Peyton et al., 1976) and are able to escape the rain and cloud filter of the troposphere due to their relatively inert chemical behaviour. Therefore, CS_2 and COS are reasonably abundant in the earth's atmosphere and are detected in the troposphere as well as in the stratosphere (e.g. Torres et al., 1980, Inn and Vedder, 1981). A nearly constant COS mixing ratio of about 500 pptv was observed throughout the whole troposphere. Both, Turco's as well as Crutzen's model predominantly consider this tropospheric COS (and CS_2) to give sufficient support to the stratospheric gaseous sulfur necessary for the aerosol formation.

Based on stratospheric in-situ aerosol measurements several authors (Junge, 1974; Hofmann et al., 1976), Lazrus and Gandrud (1974)) postulated a necessary net sulfur flux of 5.6×10^6, 2×10^7, and 9×10^6 molecules $cm^{-2}s^{-1}$, respectively. Crutzen calculated 1.8×10^7 molecules

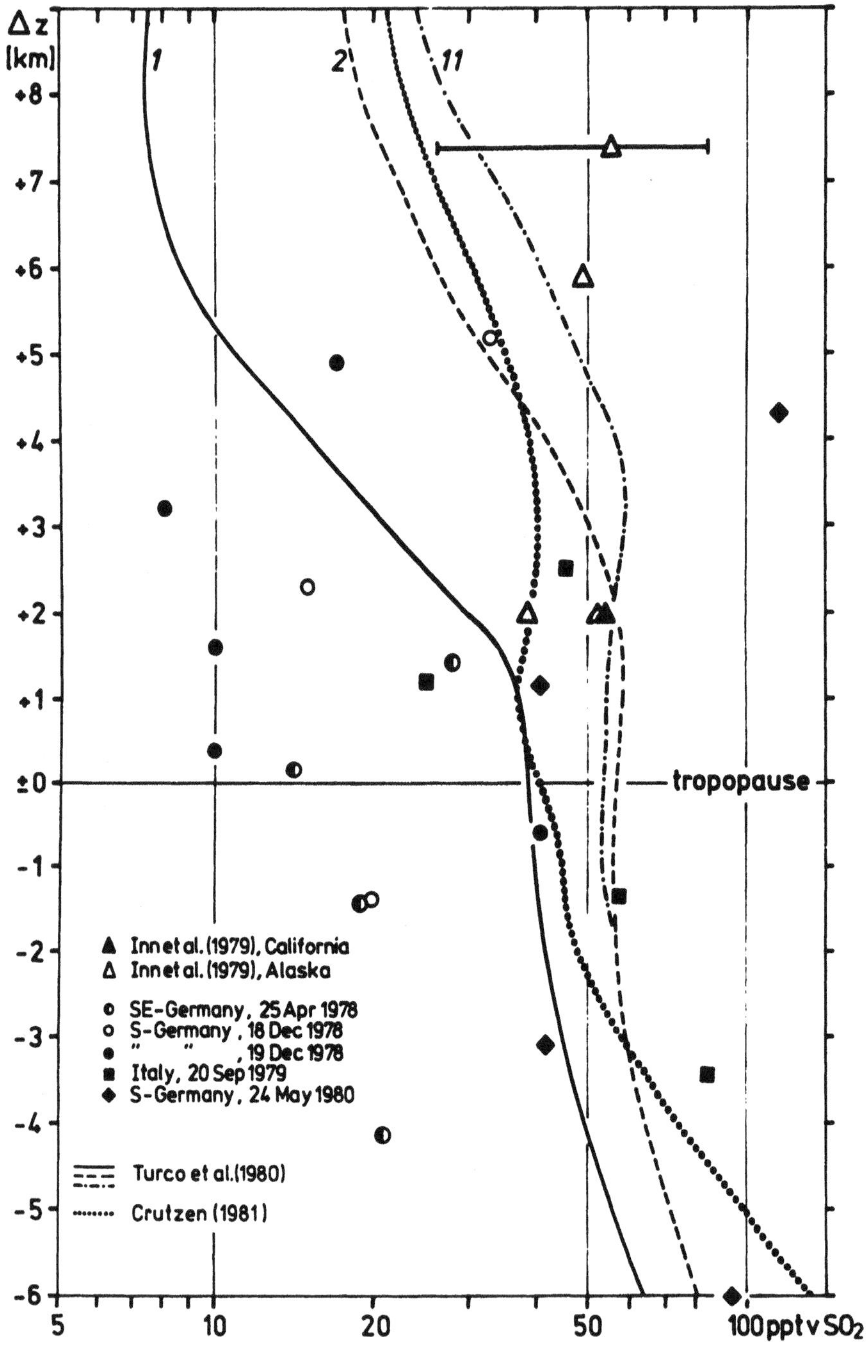

Fig. 5: *Vertical distribution of measured and calculated SO_2 mixing ratios relative to the height of the tropopause.*

Tab. 4: *Calculated fluxes of gaseous sulfur compounds across the tropopause (mol $cm^{-2}s^{-1}$)*

	Whitten et al.	Crutzen
CS_2 + COS	2.1×10^7	1.7×10^7
SO_2	1.2×10^6	4.7×10^5
H_2SO_4	3.2×10^5	---

$cm^{-2}s^{-1}$, while the calculations of Whitten et al. (1980), which are the basis for the model of Turco et al. gave 2.27×10^7 molecules $cm^{-2}s^{-1}$. Furthermore, the individual contributions of the gaseous compounds to the sulfur flux across the tropopause were also determined and are given in Tab. 4. According to these results the flux of the tropospheric SO_2 into the stratosphere seems to be of minor importance only.

This is confirmed by the measured SO_2 gradients as given in Tab. 1. Assuming that a vertical gradient of -0.5 pptv SO_2/km is also applicable to the height of the tropopause, the vertical flux of SO_2 into the stratosphere was determined adopting the K_Z-values at the tropopause used by the above mentioned models (see Tab. 5). The agreement of these SO_2 fluxes with the calculated data (see Tab. 4) is excellent.

Tab. 5: *Results of SO_2 flux calculations (details see text)*

SO_2 flux across the tropopause	effective eddy diffusion coefficient K_z at tropopause	
mol $cm^{-2}s^{-1}$	cm^2s^{-1}	reference for K_z
1.2×10^6	2.8×10^4	Turco et al. (1980)
1.5×10^6	3.4×10^4	
1.5×10^6	4.0×10^4	Crutzen (1981)

In summary it can be concluded that the flux of tropospheric SO_2, mainly man-made, is not able to maintain the stratospheric aerosol layer at times of minor volcanic activity. Nevertheless, the original concept of Junge et al. (1961) which considered the quasi-static diffusion of tropospheric, sulfur containing trace gases to be the source of the stratospheric aerosol layer, remains valid.

ACKNOWLEDGEMENTS

This work was supported by German Research Foundation, Sonderforschungsbereich 73: "Atmosphärische Spurenstoffe".

REFERENCES

BAULCH, D.L., Cox, R.A., HAMPSON JR., R.F., KERR, J.A., TROE, J., WATSON, R.T. (1980): Evaluated kinetic and photochemical data for atmospheric chemistry, J. Phys. Chem. Ref. Data, 9, (2)
CRUTZEN, P.J. (1976): The possible importance of CSO for the sulfate layer of the stratosphere, Geophys. Res. Lett., 3, 73-76
CRUTZEN, P.J. (1981): On the vertical distribution of gaseous sulfur compounds, unpublished manuscript

DANIELSEN, E.F. (1968): Stratospheric-tropospheric exchange based on radioactivity, ozone and potential vorticity, J. Atm. Sci., 25, 502-518

DANIELSEN, E.F., BLECK, R., SHEDLOVSKY, J., WARTBURG, A., HAAGENSON, P., and POLLOCK, W. (1970): Observed distribution of radioactivity, ozone and potential vorticity associated with tropopause folding, J. Geophys. Res., 75, 2353-2361

DANIELSEN, E.F. (1975): Transport by mean and turbulent motions, chapt. 6 in: The natural stratosphere of 1974, CIAP monograph 1, CIAP monograph series, U.S. Dept. of Transportation, DOT-TST-75-1

GEORGII, H.-W. and MEIXNER, F.X. (1980): Measurement of the tropospheric and stratospheric SO_2-distribution, J. Geophys. Res., 85, 7433-7438

HARKER, A.B. (1975): The formation of sulfate in the stratosphere through the gas phase oxidation of sulfur dioxide, J. Geophys. Res., 80, 3399-3401

HOFMANN, D.J., ROSEN, J.M., PEPIN, T.J. and PINNICK, R.G. (1976): Stratospheric aerosol measurement IV: Global time variations of the aerosol burden and source considerations, J. Atm. Sci., 33, 1782-1788

INN, E.C.Y. and VEDDER J.F. (1981): Measurement of stratospheric sulfur constituents, Geophys. Res. Lett., 8, 5-8

IYER, R.S. and ROWLAND, F.S. (1980): A significant upper limit for the rate of formation of OCS from the reaction of OH with CS_2, Geophys. Res. Lett., 7, 797-800

KURYLO, M.J. (1978): Flash photolysis resonance fluorescence investigation of the reactions of OH radicals with OCS and CS_2, Chem. Phys. Lett., 58, 238-242

JAESCHKE, W., GEORGII, H.-W. and SCHMITT, R. (1976): Preliminary results of stratospheric SO_2 measurements, Geophys. Res. Lett., 3, 517-519

JUNGE, C.E., CHAGNON, C.W. and MANSON, J.E. (1961): Stratospheric aerosols, J. Met., 18, 81-108

JUNGE, C.E. (1974): Sulfur budget of the stratospheric aerosol layer, Proc. of the IAMAP Conf. on Structure, Composition and General Circulation of the Upper and Lower Atmospheres, Melbourne, Jan. 1974, Vol. 1, pp. 85-97

LAZRUS, A.L. and GANDRUD, B.W. (1974): Stratospheric sulfate aerosol, J. Geophys. Res., 79, 3424-3431

MEIXNER, F.X. (1981): Die vertikale Verteilung des atmosphärischen Schwefeldioxids im Tropopausenbereich, PhD thesis, J.W. Goethe University, Frankfurt am Main, FRG

MEIXNER, F.X., GEORGII, H.-W., OCKELMANN, G., JÄGER, H. and REITER, R. (1981): The arrival of the Mount St. Helens eruption cloud over Europe, Geophys. Res. Lett., 8, 163-166

MOORTGART, G.J. and JUNGE, C. (1977): The role of SO_2 oxidation for the background stratospheric sulfate layer in the light of new reaction data, PAGEOPH, 115, 769-774

NASA (1981): Chemical kinetic and photochemical data for use in stratospheric modeling evaluation, No. 4, NASA panel for data evaluation, Jan. 15, 1981, NASA Jet Prop. Lab., Calif. Inst. Technology, Pasadena, Cal.

RAVINSHANKRA, A.R., KREUZZER, N.M. SHAH, R.C., WINE, P.H. (1980): Rate of reaction of OH with COS, EOS, 61, 671

PEYTON, T.O., STEELE, R.V. and MABEY, W.R. (1976): Carbon disulfide, carbonyl sulfide: Literature review and environmental assessment, Standford Research Institute Report 68-01-2940. 57 pp.
REITER, E.R. (1975): Stratospheric-tropospheric exchange processes, Rev. Geophys. Space Phys., 13, 459-474
SHAPIRO, M.A., REITER, E.R., CADLE, R.D., and SEDLACEK, W.A. (1980): Veertical mass- and trace constituent transports in the vicinity of jet streams, Arch. Met. Geophys. Biokl., Ser. B., 28, 193-206
STAUFF, J. and JAESCHKE, W. (1975): A chemiluminescence technique for measuring atmospheric trace concentrations of SO_2, Atm. Environ., 9, 1038-1039
TORRES, A.L., MAROULIS, P.J., GOLDBERG, A.B. and BANDY, A.R. (1980): Atmospheric OCS measurements on Project Gametag, J. Geophys. Res., 85, 7357-7360
TURCO, R.P., WHITTEN, R.C., TOON, O.B., INN, E.C.Y., and HAMILL, P. (1980): Stratospheric hydroxyl radical concentrations: New limitations suggested by observations of gaseous and particulate sulfur, J. Atm. Sci., in press
WHITTEN, R.C., TOON, O.B. and TURCO, R.P. (1980): The stratospheric aerosol layer: Processes, models, observations, and simulations, PAGEOPH, 118, 87-127

The Photolysis of Acetaldehyde Under Atmospheric Conditions

H. Meyrahn, G. K. Moortgat and P. Warneck

Max-Planck-Institut für Chemie, Air Chemistry Division, Saarstraße 23, Mainz, FRG

A b s t r a c t

The photodecomposition of mixtures of 100 - 200 ppm CH_3CHO in 1 atm of air or purified N_2 has been studied at 300 K and the quantum yields of CO and CH_4 determined in the spectral range 250 - 340 nm. The quantum yields are nearly identical in both systems. $\emptyset_{CO}$ increases monotonically from zero at 330 nm to near 0.8 at 250 nm. $\emptyset_{CH_4}$ remains near zero at $\lambda > 285$ nm and increases sharply to 0.5 at 250 nm. A pressure dependence of $\emptyset_{CO}$ was observed at 313 nm, indicating that some quenching occurs. No H_2 was observed. Quantum yields for both primary steps

$$CH_3CHO \longrightarrow CH_3 + CHO \qquad \emptyset_1$$
$$\longrightarrow CH_4 + CO \qquad \emptyset_2$$

are derived. Measured CO production rates in sunlit CH_3CHO/Air and N_2 mixtures produce noontime dissociation rates of 3×10^{-6} sec^{-1} at 30° zenith angle. This is a magnitude lower as the currently accepted values of Meagher et al.(1976).

1. INTRODUCTION

The photooxydation of acetaldehyde in the atmosphere can occur by two routes: one is direct photodecomposition, the other is a reaction with OH which abstracts the aldehydic hydrogen and leads to the formation of acetyl radicals. Then they add oxygen to form peroxyacetyl which is the precursor of peroxyacetylnitrate (PAN), an air pollutant having important physiological effects. PAN probably occurs also naturally in small concentrations. The relative contributions of the two oxidation pathways are critical for an assessment of PAN formation in the natural atmosphere, but have remained uncertain because photodecomposition quantum yields for acetaldehyde under natural conditions had remained undetermined. We have measured CO and CH_4 product quantum yields in mixtures of about 100 ppm of acetaldehyde in air and report the results here. It can be shown that under these experimental conditions acetaldehyde is not subject to secondary attack by radicals formed in the photolysis.

2. EXPERIMENTAL

Two Xe arc monochromator set-ups (A,B) with different light throughput (A= Jobin-Yvon, f/2 and B= Spex, f/4) were used. Mixtures of acetaldehyde in air or N_2 were prepared in a Hg and grease-free handling system and then expanded simultaneously into 3 identical quartz gas cells (25 cm³, i.d. = 18 mm): one was used for the determination of the CO/CH_4 background, one served as blank cell (kept in dark during the photolysis), and one as the photolysis cell proper. A calibrated thermopyle and/or acetone actinometry ($\emptyset_{CO}$=1 at 125°C) was used to measure the photon fluxes. In the analysis CO was converted to CH_4 by means of a Ni - catalyst. All the gas samples were analysed gas chromatographically.

3. QUANTUM YIELDS

It made no difference in the quantum yields observed whether irradiations were performed with mixtures of acetaldehyde in air or nitrogen. Although prepurified nitrogen was used, its oxygen content probably was still about 1 ppmv which would not have been low enough to change the prevalent reaction sequence. Figure 1 shows the variation of CO and CH_4 quantum yields versus wavelength with the total pressure fixed at one atmosphere. $\emptyset_{CO}$ increases monotonically from zero at 330 nm to near 0,8 at 250 nm. $\emptyset_{CH_4}$ remains near zero at wavelengths greater than 285 nm, then increases sharply towards shorter wavelengths. The threshold energy corresponds to 420 KJ/mole. Within experimental error there was no dependence on the quantum yield with light intensities varying within a factor of four (5.10^{13} - 2.10^{14} photons/cm²s). Hydrogen was not found in significant amounts, indicating that the primary process

$$CH_3CHO + h\nu \longrightarrow CH_2CO + H_2$$

does not occur. Also Blacet and coworkers (1942) working with pure CH_3CHO did not detect H_2 as a photolysis product. Consequently, also

$$CH_3CHO + h\nu \longrightarrow CH_3CO + H$$

cannot be an important process. CO_2 was not detected, and other products were not searched for. Therefore, under our experimental conditions, the main reactions are

$$CH_3CHO + h\nu \longrightarrow CH_3 + CHO \qquad (1)$$

$$CH_3CHO + h\nu \longrightarrow CH_4 + CO \qquad (\lambda < 285) \qquad (2)$$

followed by

$$CHO + O_2 \longrightarrow HO_2 + CO \qquad (3)$$

$$CH_3 + O_2 + M \longrightarrow CH_3O_2 + M \qquad (4)$$

On the assumption that no CO nor CH_4 are produced from secondary reactions, the primary quantum yields at atmospheric pressure are given by

$$\emptyset_1 = \emptyset_{CO} - \emptyset_{CH_4}$$

$$\emptyset_2 = \emptyset_{CH_4}$$

and are displayed in Fig.2.

Experiments at 313 nm indicate that the CO quantum yields both in air and N_2 mixtures increase as the total pressure is lowered. This behavior is still under investigation.

4. PHOTODISSOCIATION RATES

The quantum yield data can be readily applied to the troposphere. At ground level, radiation at wavelengths below 295 nm is absent due to the filtering effect of stratospheric ozone. As a consequence, the production of methane from the photodecomposition of acetaldehyde can be neglected. According to our present knowledge, CO appears to be the only stable product resulting from the processes (1) and (3). The corresponding acetaldehyde photodissociation and CO production rates are then identical.

Photodissociation coefficients may be calculated from

$$J_{CH_3CHO} = \int \phi_{CO}(\lambda)\,\sigma(\lambda)\,I_0(\lambda)\,d\lambda$$

where $\phi_{CO}(\lambda)$ and $\sigma(\lambda)$ are the quantum yield and absorption cross section, respectively, both as a function of wavelength as determined by us. $I_0(\lambda)$ is the solar radiation intensity as a function of zenith angle. We have used data from Peterson (1976) for middle latitudes. The integral was evaluated and is shown in Fig. 3. The calculated photoproduction rates were further compared with experimental values obtained from the acetaldehyde photodecomposition in sunlight. For this purpose several quartz bulbs filled with mixtures of 100 ppm CH_3CHO in air or N_2 were exposed to sunlight during noon hours in July 9-28 1981. The photodissociation coefficients derived from the production rates are also shown in Fig. 3 and give an average value $J_{CH_3CHO} = 3 \cdot 10^{-6}\ s^{-1}$.

5. RELATIVE IMPORTANCE OF ACETALDEHYDE PHOTODECOMPOSITION IN THE ATMOSPHERE

All aldehydes react very rapidly with OH radicals. The measured rate coefficient of the reaction with acetaldehyde is $k = 1.6 \cdot 10^{-11}\ cm^3/molecule \cdot s$ (Atkinson and Pitts 1978). If we assume a noon value for the OH concentration of 10^6 molecules/cm³, one obtains a removal rate for acetaldehyde due to reaction with OH of $1.6 \cdot 10^{-5}\ s^{-1}$, which is about five times greater than that due to photolysis. Weaver et al. (1976) assumed additional CO formation due to chemical quenching of triplet acetaldehyde and deduced a photodissociation coefficient of $1.2 \cdot 10^{-5}\ s^{-1}$, which is about equal to the rate of removal by OH. Our results suggest that acetaldehyde is predominantly removed by reaction with OH whereas photolysis amounts to less than 20% of the total. Accordingly, the acetaldehyde in the lower troposphere is predominantly converted to acetylperoxyradicals leading to the subsequent formation of PAN.

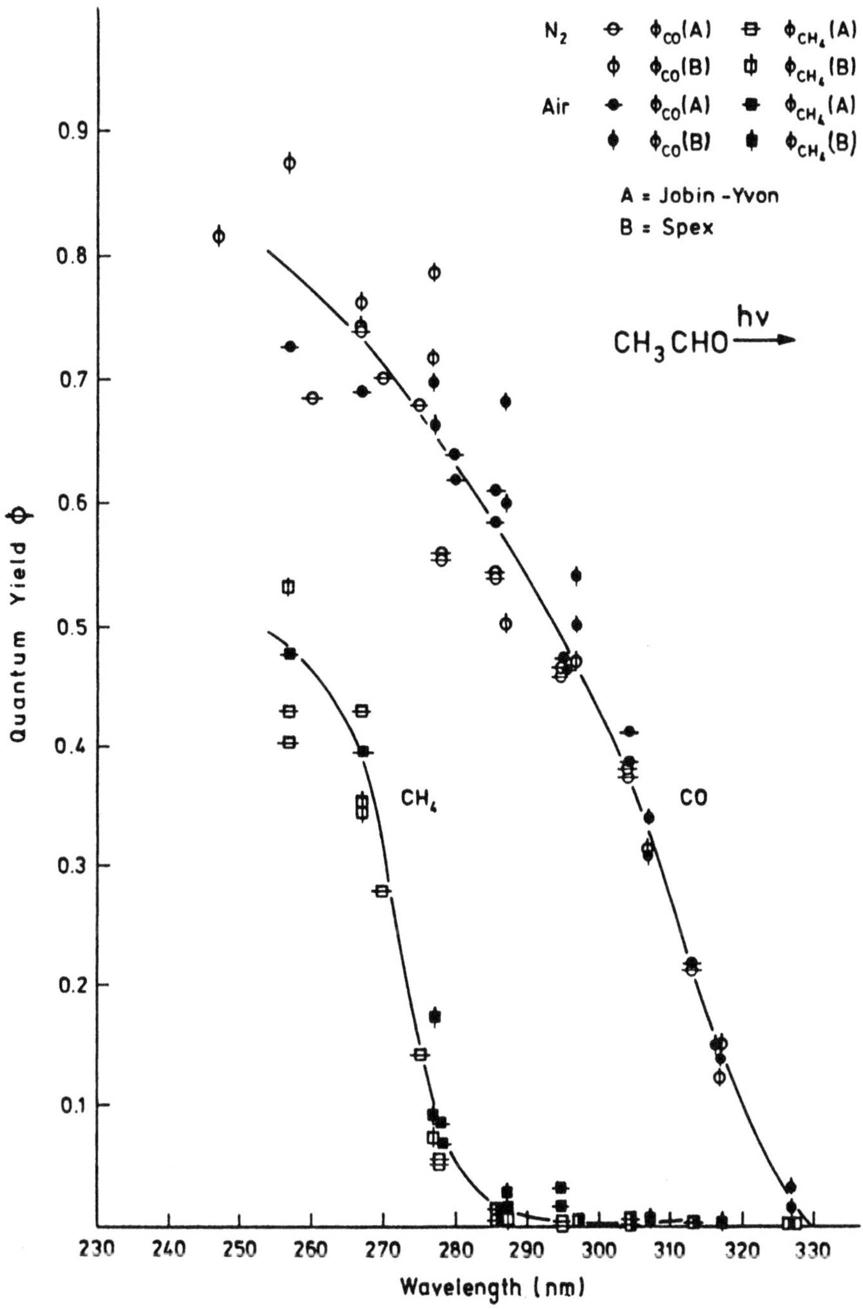

Figure 1: CO and CH_4 quantum yields as a function of wavelength (about 100 ppmv acetaldehyde in air and N_2 at 1 atm)

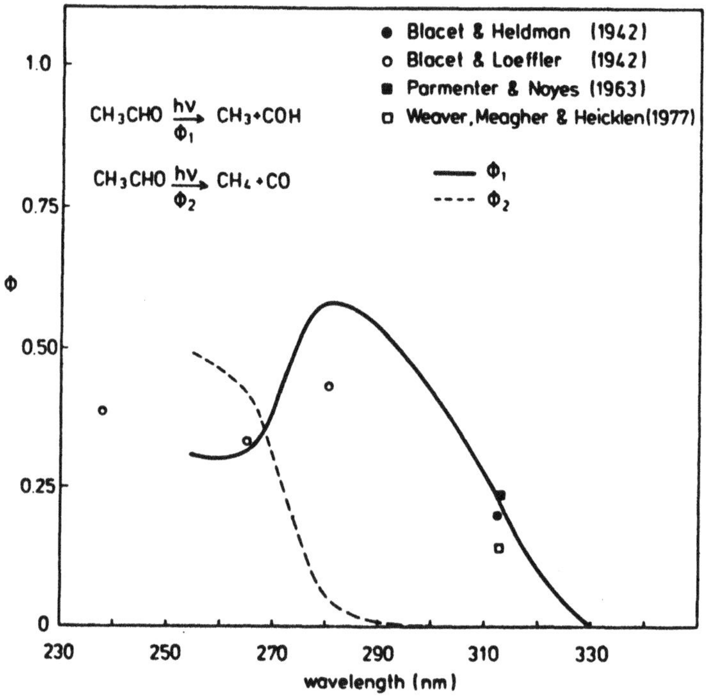

Figure 2: Primary quantum yields at atmospheric pressure

$$\emptyset_1 = \emptyset_{CO} - \emptyset_{CH_4}$$

$$\emptyset_2 = \emptyset_{CH_4}$$

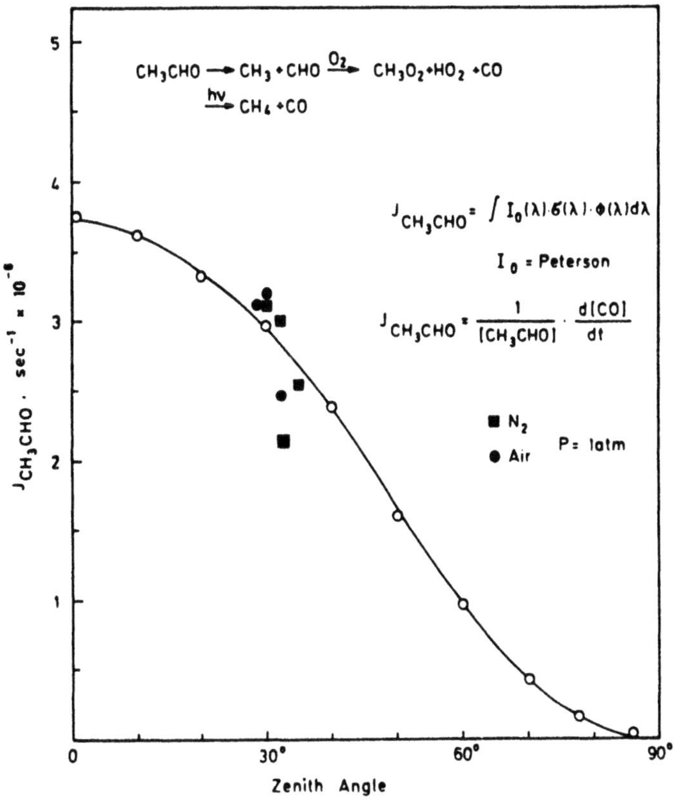

Figure 3: Calculated (o) and measured (● , ■) acetaldehyde photodissociation coefficients as a function of zenith angle

6. REFERENCES:

ATKINSON, R. and PITTS, Jr. J.N., (1978) Kinetics of the reactions of the OH radical with HCHO and CH_3CHO over the temperature range 299-426 K.,J. Chem. Phys. 68, 3581 - 3584.

BLACET, F.E. and HELDMAN, J. D. (1942) The photolysis of aliphatic aldehydes X. Acetaldehyde and iodine mixtures. J. Am. Chem. Soc. 64, 889 - 893.

BLACET, F.E. and LOEFFLER, D.E. (1942) The photolysis of aliphatic aldehydes XI. Acetaldehyde and iodine mixtures. J. Am. Chem. Soc. 64, 893 - 896.

PARMENTER, C.S. and NOYES Jr. W.A., (1963) Energy dissipation from excited acetaldehyde molecules. J. Am. Chem. Soc. 85, 416 - 421.

PETERSON, J. T., (1976) Calculated actinic fluxes (290 - 700 nm) for air pollution photochemistry applications. Rept. EPA-600/4-76-025. U.S. Environ. Prot. Agency. Research Triangle Parl, N.C. (NTIS PB 255819)

WEAVER, J., MEAGHER, J. and HEICKLEN, J., (1976/77) Photo-oxidation of CH_3CHO vapor at 3130 Å., J. Photochem., 6, 111 - 126.

Tropospheric OH Measurements

G. Ortgies and F. J. Comes

Institut für Physikalische und Theoretische Chemie der J. W. Goethe-Universität Frankfurt/Main, FRG

A b s t r a c t

Laser-optical methods for determining tropospheric OH concentrations are based on the absorption of resonance radiation by the OH radicals. The interfering effects, saturation and self-generating, which are encountered in the method of laser-induced fluorescence (LIF) are discussed with respect to the applicability of this method. An absorption method is introduced as an alternative. An interference effect due to SO_2 is described and experimental results for tropospheric OH and SO_2 are presented.

1. Introduction

To understand the chemistry of the atmosphere, it is necessary to know both the concentrations of various atmospheric trace gases and the rate constants of many elementary chemical processes. A central role in this chemistry is played by the OH radical because of its very fast reaction with nearly all trace gases. The main atmospheric OH source is the reaction of water with excited oxygen atoms ($O(^1D)$) which are generated in the photolysis of ozone by sunlight at wavelengths below 310 nm. Due to the high reactivity of OH, its lifetime is small (~ 1 s) under tropospheric conditions and its average concentration extremely low. Model calculations (Warneck (1975), Crutzen and Fishman (1977)) predict concentrations of the order of 10^6 cm^{-3}. To determine in situ tropospheric OH concentrations, chemical and optical measuring methods have been employed. Until now, the lowest concentrations have been determined with the chemical method (Campbell et al., 1979), which is based on measurement of the rate of oxidation of CO to CO_2 by OH. However, as this method is subject to several uncertainties, it should be verified by an optical method. In recent years measurements have been carried out with laser-optical methods by Wang and Davis (1974), Wang et al. (1975), and Davis et al. (1976, 1979). Their measurements with the laser-induced fluorescence technique are afflicted by a severe interference problem, as shown by Ortgies et al. (1980, 1981) and Gericke and Comes (submitted for publication). The limitations imposed on the use of this technique by saturation and OH self-generating will be described in the following section. A laser-optical method will be introduced as an alternative measuring method.

2. Principle of the LIF method

The LIF method is shown schematically in Fig. 1. From the ground state ($^2\Pi$,v"=0) electrons are pumped by the laser radiation (ν_{ABS},λ=282 nm) to the first excited state ($^2\Sigma^+$,v'=1). This state is depopulated by vibrational relaxation (σ_{10}), quenching (σ_1) and spontaneous emission (ν_1) to $^2\Pi$,v"=1.

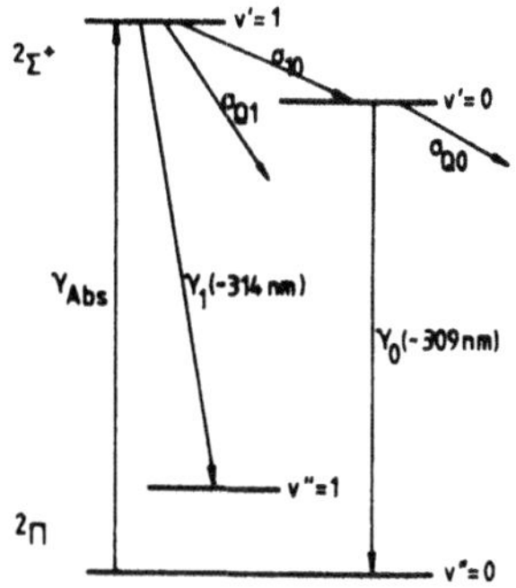

Fig. 1. OH level diagram describing the LIF technique schematically. σ_0 and σ_1 are quenching cross sections. ν_0 and ν_1 relate to the free molecule lifetimes. ν_{ABS} is the absorption rate.

Of these processes the vibrational relaxation is the most effective under tropospheric conditions, so that the fluorescence radiation (ν_0,λ= 309 nm), which is shifted to longer wavelengths relative to the exciting laser radiation, can be observed.

However, this measuring method suffers from a severe interference problem. The exciting laser radiation at 282 nm dissociates the tropospheric ozone via

$$O_3 + h\nu(282\ \mathrm{nm}) \rightarrow O(^1D) + O_2(^1\Delta_g) \qquad (1)$$

The $O(^1D)$-atoms react rapidly with water to form two OH radicals:

$$O(^1D) + H_2O \rightarrow OH + OH \qquad (2)$$

Competing processes are:

$$O(^1D) + M \rightarrow O(^3P) + M \ ; \ M = N_2, O_2 \qquad (3)$$

At ground level the OH radicals are formed in less than 1 ns. The laser pulse lengths hitherto applied to the LIF method were $\Delta T = 7$ ns. Therefore it is possible to produce OH radicals and excite them with the same laser pulse. Thus the signal obtained by the LIF method is the sum of the fluorescence signal of the natural OH and the fluorescence signal of the self-generated OH. To calculate the concentration of the self-generated $[OH]_{gen}$, eq. (4) must be solved:

$$[OH]_{gen} = \int_0^{\Delta T} [OH] \cdot n(t) \cdot dt \ / \int_0^{\Delta T} n(t) \cdot dt \qquad (4)$$

In this equation $[OH]$ stands for the concentration of OH formed, via eq. (2), in the lower rovibronic state. The photon flux of the laser radiation is described by n(t). These calculations were carried out in detail by Ortgies et al. (1980, 1981) and were later confirmed by Davis et al. (1981). The results of these calculations are presented in Fig. 2 for laser pulse lengths of $\Delta T = 7$ ns (upper diagram) and $\Delta T = 1$ns (lower diagram). In each diagram the curve b) represents the case where, for the given exciting laser energy density and altitude, 1×10^6 OH cm^{-3} are produced and "seen". In the hatched region below these curves the self-generation is even greater than 1×10^6 cm^{-3}. Recent results (Gericke and Comes, 1982; Gericke and Comes, submitted for publication) show that the effect of self-generation is increased by a factor

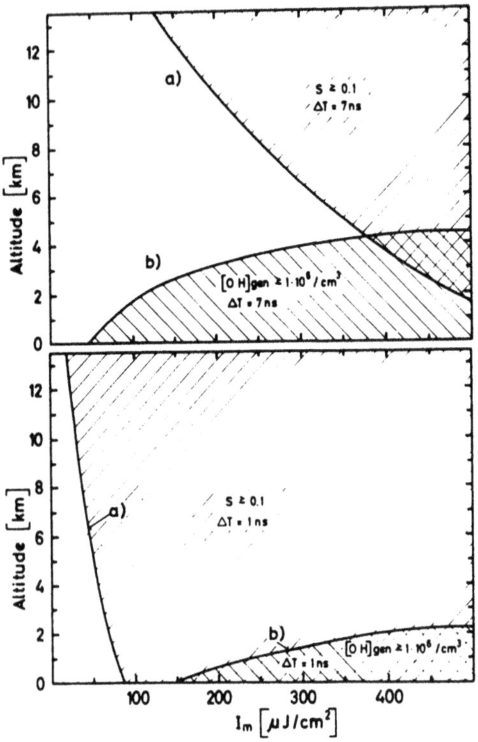

Fig. 2a, b. The diagram shows the feasible range of laser intensity in measurements of tropospheric OH with the LIF method as function of altitude for laser pulse lengths a) ΔT = 7 ns, b) ΔT = 1 ns. In both parts of the figure the shaded area above curve a) is forbidden due to saturation. The shaded area below curve b) is forbidden due to a growth of $[OH]_{gen}$ from a value of 1×10^6 cm^{-3}.

of two at ground level due to rotational relaxation of the initial rotational distribution of OH, formed in reaction (2).

Quantitative fluorescence measurements presuppose that the absorption rate varies linearly with the exciting laser intensity. Linear dependence is not realized for high intensities. This saturation effect has been discussed in detail by Killinger et al. (1976) and Ortgies et al. (1981). The results of these calculations are also presented in Fig. 2. In each diagram the curve a) represents the case of a 10 % deviation from linear absorption. In the hatched region below these curves the saturation effect is even greater. The saturation parameter S, the ratio of pump rate and average relaxation rate, depends strongly on the altitude, especially for short laser pulse lengths due to the increasing relaxation time with increasing altitude. The advantage of decreasing self-generation with shorter pulse lengths is canceled by the increasing saturation. The results of these considerations, which are summarized in Fig. 2, show that the tropospheric OH data collected with the LIF method are subject to great doubts.

3. The long path absorption method

A method which excludes the above-discussed interference effects, is the long-path absorption method. Due to the very low tropospheric OH concentration very long absorption paths are necessary in spite of the relatively high transition probabilities of OH in the uv. To obtain absorptions between 0.01 % and 0.1 % the length of the measuring path must be several kilometers. For intensity reasons lasers are the usual light sources for such long absorption paths. Another reason for laser use is the need for high energy densities at the resonance wavelength ($\lambda \sim 308$ nm). In Fig. 3 the experimental setup is drawn schematically.

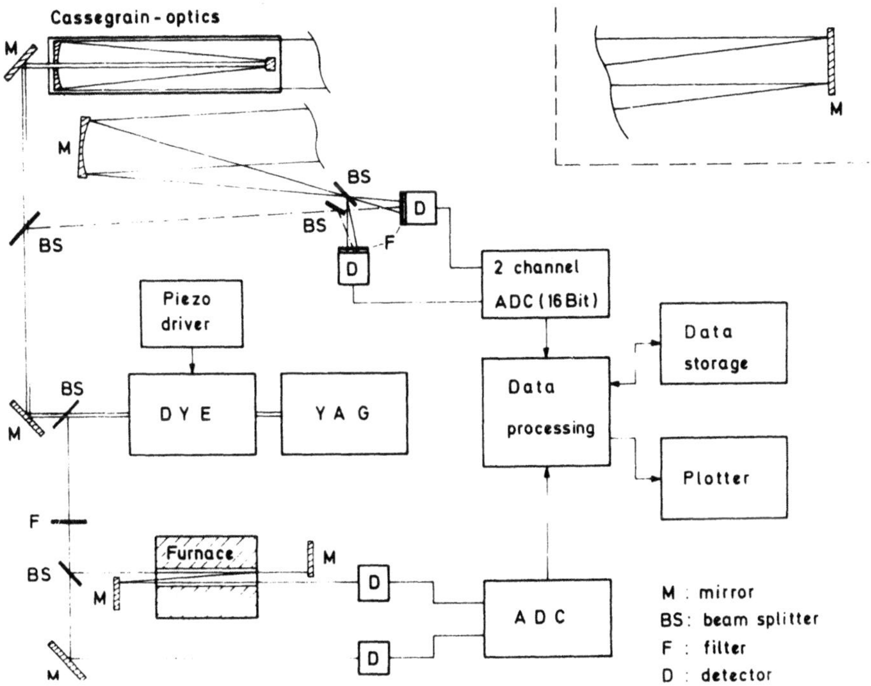

Fig. 3. Schematic diagram of the experimental arrangement.

A frequency-doubled Nd-YAG pumped tunable dye laser is used as light source. To reduce the divergence and the energy density of the radiation, the laser beam is widened by a Cassegrain telescope. Ultraviolet and visible laser radiation are sent to a plane mirror stationed at a distance of 3.3 km, from where it is reflected back to the laboratory and focused on detectors by means of a concave mirror. To elimimate intensity fluctuations of the laser, part of the radiation is brought out to the same detectors. The use of the same detectors to avoid faults due to different characteristics is possible, because the pulse which traverses the absorption path is delayed by about 20 µs. The visible and uv radiation is separated by filters. The analog signals are digitized and received by a MINC-11 computer, where they are stored and evaluated. The frequency calibration is also done by an absorption measurement, as shown in the lower part of Fig. 3. The laser wavelength is alternately tuned to resonance with the OH line and off resonance ($\Delta\nu = 2\ cm^{-1}$) from pulse to pulse by means of a piezo ceramic. Thus the method becomes an absolute measuring method. For these measurements the $Q_1(2)$- transition of OH ($^2\Pi, v''=0 \rightarrow {}^2\Sigma^+, v'=0$) is used. This is the strongest transition under tropospheric conditions. To achieve high absorptions, the laser bandwidth ($\Delta\nu_L=0.06\ cm^{-1}$) was chosen to be smaller than the linewidth of the OH resonance transition ($\Delta\nu_{OH}=0.15\ cm^{-1}$).

With measurements of such low concentrations there always exists the danger of interference effects. In the case of the absorption method an interference can occur if the background absorption differs for the different wavelengths (on line, off line). One trace gas which absorbs in the uv and is found in the troposphere in high concentrations is SO_2. To estimate a possible interference effect, the SO_2 spectrum has been obtained in the 308 nm spectral region. The result is presented in Fig. 4. The influence of pressure broadening can be clearly realized by changing the pressure from 100 torr to 400 torr. The wavelength positions for the OH measurement are

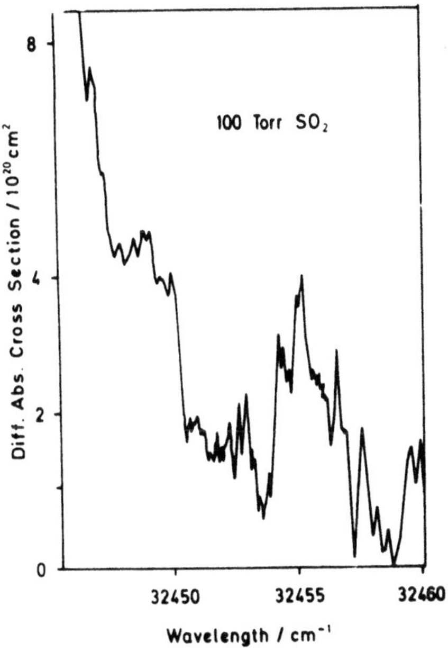

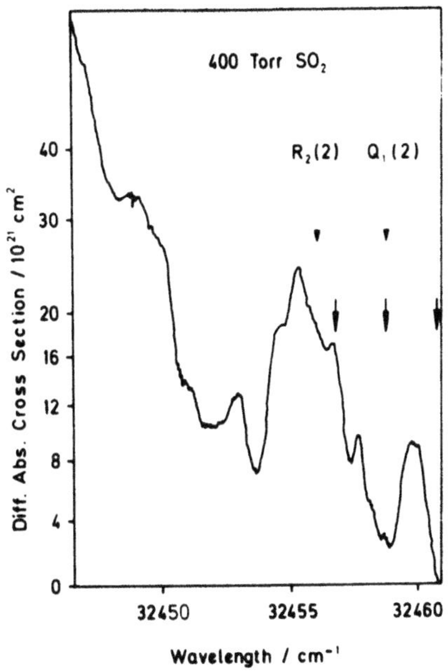

Fig. 4. Absorption spectrum of SO_2 in the 308 nm spectral region.

marked by the three arrows. Fig. 4 shows that the absorption cross section changes in this narrow spectral range. Thus, depending on the SO_2 partial pressure, there can be an interference effect and using the absorption method necessitates determining the SO_2 concentration, too. This can be easily done by slightly changing the described method of measurement (Ortgies and Comes, to be published).

4. Results

With the absorption method described above, measurements were carried out in Frankfurt. The signal which traverses the absorption path is attenuated by wavelength-dependent absorption due to atmospheric trace gases, Rayleigh scattering, Raman scattering, and scattering by dust and aerosols. Depending on meteorological parameters the transmitted signal varies between 0.1 % and 5 % of the incident signal. Fluctuations of the particle density in the atmosphere give rise to strong fluctuations of the signals. Whereas in the LIF method the signal-to-noise ratio is mainly determined by the value of the signal, in the absorption method it depends on the signal fluctuations. These fluctuations limit an atmospheric absorption measurement to absorptions of 0.34 % averaging over 10 minutes. This corresponds to a detection limit of 2×10^7 OH cm^{-3}.

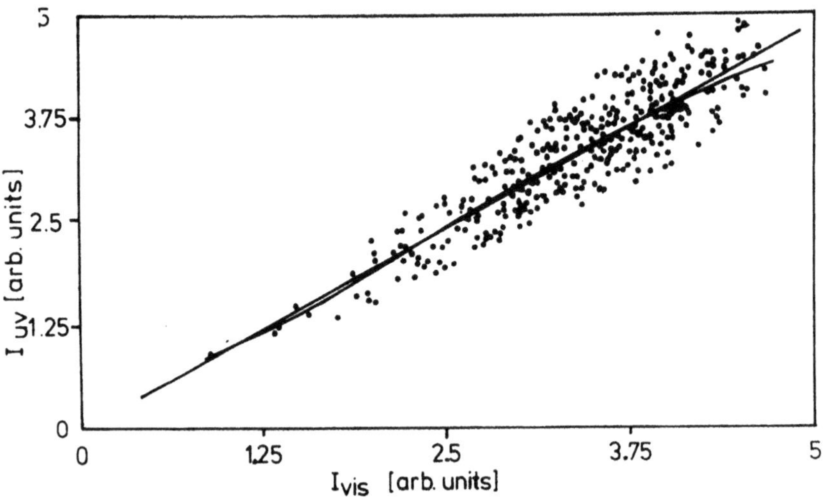

Fig. 5. Correlation diagram of the visible and uv laser radiation. The solid lines represent a first order and third order least squares fit, respectively.

The advantage of this method is based on the fact that both the visible and the uv radiation of the dye laser traverse the absorption path. In Fig. 5 the very good correlation of visible and uv radiation of the same laser pulse is presented. Using this correlation for the evaluation, the detection limit can be improved to 4×10^6 cm^{-3} for the same averaging time. Measurements in the summer of 1981 show that the OH concentrations were below this detection limit. At the same time SO_2 measurements were carried out with the long path absorption technique. SO_2 concentrations could be detected for various meteorological conditions in the range of

2 ppb and 40 ppb with a precision of 0.5 ppb. Future measurements of OH and SO_2 in a clean-air region are planned. To verify the results, parallel measurements will also be carried out with a continous-absorption method (Perner et al. 1976).

References

Campbell, M.J., J.C. Sheppard, and B.F. Au, Geophys. Res. Lett. 3, 175 (1979)

Crutzen, P.J. and J. Fishman, Geophys. Res. Lett. 4, 321 (1977)

Davis, D.D., W. Heaps, and T. McGee, Geophys. Res. Lett. 3, 331 (1976)

Davis, D.D., W. Heaps, D. Philen, and T. McGee, Atm. Environ. 13, 1197 (1979)

Davis, D.D., M.O. Rodgers, and S.D. Fischer, Geophys. Res. Lett. 8, 73 (1981)

Gericke, K.-H., and F.J. Comes, Chem. Phys. 65, 113 (1982)

Gericke, K.-H., and F.J. Comes, (submitted for publication)

Killinger, D.K., C.C. Wang, and U. Hanabusa, Phys. Rev. A 13, 2145 (1976)

Ortgies, G., K.-H. Gericke, and F.J. Comes, Geophys. Res. Lett. 7, 905 (1980)

Ortgies, G., K.-H. Gericke, and F.J. Comes, Z. Naturforsch. 36a, 177 (1981)

Ortgies, G., and F.J. Comes, (to be published)

Perner, D., D.H. Ehhalt, H.W. Patz, U. Platt, E.P. Roth, and A. Volz, Geophys. Res. Lett. 3, 466 (1976)

Wang, C.C., and L.I. Davis, Phys. Rev. Lett. 32, 349 (1974)

Wang, C.C., L.I. Davis, C.H. Wu, S. Japar, H. Niki, and B. Weinstock, Science 189, 797 (1975)

Warneck, P., Planet. Space Sci. 23, 1507 (1975)

Session II:
Aerosol Particles

Development of a New Sunphotometer and Establishment of a Network Across Africa for the Investigation of the Transport of Sahara Dust

Guillaume A. d'Almeida

Institut für Meteorologie der Universität Mainz, FRG

S u m m a r y

The matter of this project is the investigation of the large scale south transport of Saharan dust originated by critical weather conditions such as sandstorm and dustwind. A new sensitive sunphotometer has been developed for this purpose. With ten instruments a network has been installed southerly of the Sahara across Africa. It is the aim to determine the main sources of the mineral dust and its further transport to the South.

1. INTRODUCTION

Despite of its relative large uncertainty, monitoring of atmospheric turbidity by satellite is a suitable method because of its area coverage. Atmospheric turbidity remote sensing by satellite over the ocean has been carried out by several authors such as CARLSON and WENDLING (1977), KOEPKE and QUENZEL (1979), KOEPKE et al (1980), NORTON et al (1980). Investigations over land surfaces however are still inadequate because the reflection properties of land surfaces are anisotropic, the albedo is high and varies strongly with space and time (KOEPKE et al, 1980). Nevertheless, satellite images remain an additional instrument for localising sources and tracing of transport of dust. The sunphotometry is a convenient method for the determination of atmospheric turbidity over land surface. There is a large number of radiation measuring instruments with different measuring principles. The first ones have been constructed during the first half of the last century (after FOITZIG and HINZPETER, 1958) and were based on a calorimetrical measuring principle. One distinguishes the sphere thermometer after LESLIE, the pyranometer after POUILLET and the radiation measuring instrument after BELLANI. The sun measurement instruments which follow derive their principles from the theory of the thermodynamics and electricity and enable the determination of the total intensity of the sun radiation. The optical method for the determination of solar radiation came into use with the development of optical filters. According to QUENZEL (1966), FOWLE was the first to develop a spectrograph with an optical determination of water vapor at the beginning of this century. With the so-called "Potsdamer Normalfilter" (OG1: yellow filter, RG2 or RG8: red filter) actinometers were built to measure the sun intensity. With the availability of interference filters it became possible to select narrow spectral ranges of the sun radiation. VOLZ (1961) was one of the first to develop a sunphotometer with interference filters. Another sunphotometer which can be considered as prototype, was improved by FRÖHLICH (1977). The sunphotometer after FRÖHLICH (1977) has only one interference filter with 500 nm wavelength. The instrument after VOLZ (1961) has several channels, is very simple and easy to handle, but shows some limitations:

a) The instrument after VOLZ (1961) is not sensitive enough to measure under heavy turbid atmospheric conditions as it often occurs in the desert during sandstorm and dustwind weather situations. This statement is based on own observations in Matam (Senegal) and Bamako (Mali) in Februar and March 1979.

b) The determination of the air mass is quite inaccurate and should better be calculated from the sun position. Furthermore the measurement of the air mass over sea is quite impossible (VOLZ, 1961).

c) In the regions of the earth which are affected by large air humidity the unprotected interference filters and the photodiode decay quickly and cause uncontrolled errors.

It was our aim to develop an instrument of the same simplicity as VOLZ's sunphotometer, but overcome its limitations. The instrument should be robust enough to be used under rough conditions and sensitive enough to measure large turbidities.

2. DESIGN OF THE NEW SUNPHOTOMETER

The new sunphotometer will be briefly described here. A later publication will give more details. The instrument contains:

* 9 interference filters imbedded in a filter wheel and protected against deterioration. The center wavelengths of the quasi monochromatic filters are selected from 400 nm to 1025 nm with a bandwidth variing from 12 nm to 19 nm (see Fig. 1)

* A diaphragma after the filter wheel to limit the incident sun radiation and to suppress the parasitic stray light.

* A black tube between the diaphragma and the photodetector to exclude any light in the surrounding of the photodetector.

* The photodetector as photovoltaic silicon detector with a spectral sensitivity shown in Fig. 1.

* A potentiometer with a pendulum for the measurement of the solar elevation angle.

* An electronic unit with a digital and an analog part.

* Two batteries which provide the necessary energy.

* A small container with silica gel to keep the inside of the instrument dry. The filter and the detector can accordingly be protected from the humidity.

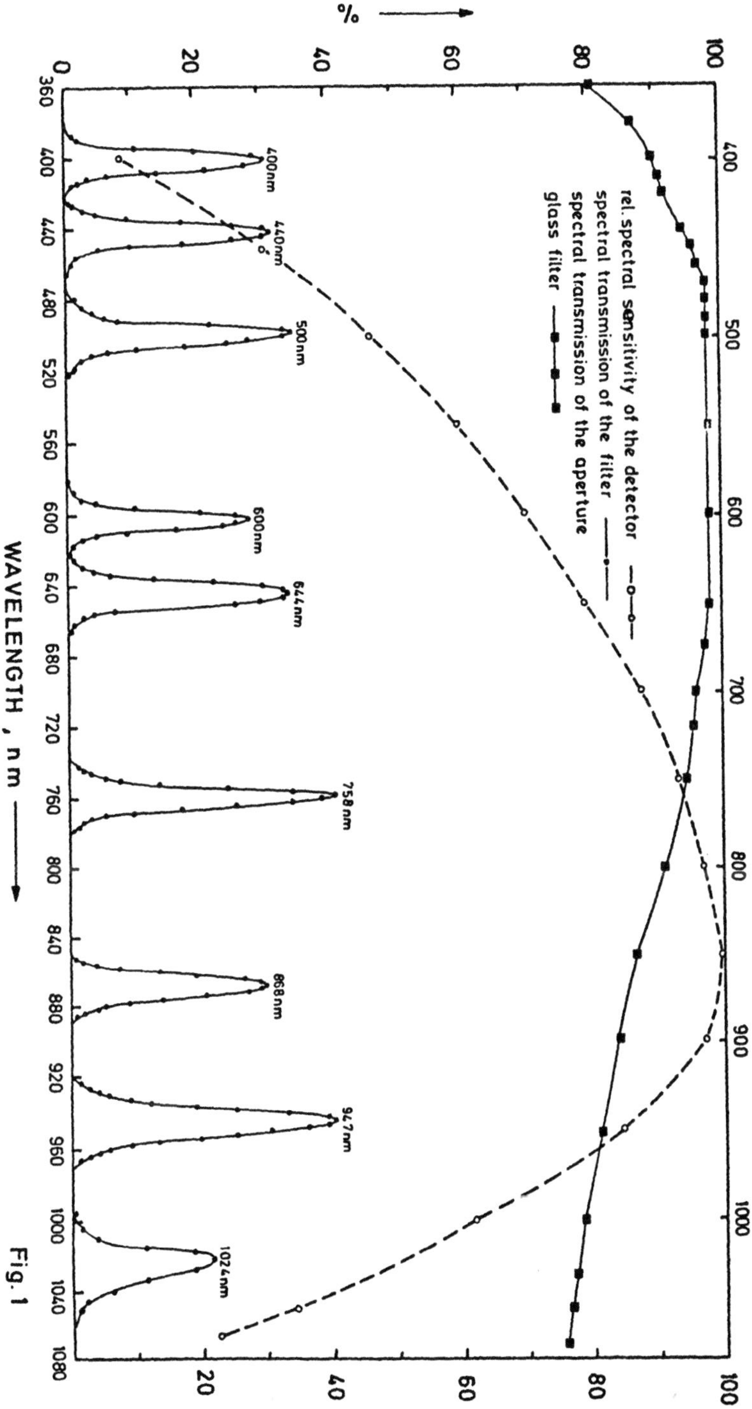
rel. spectral sensitivity of the detector
spectral transmission of the filter
spectral transmission of the aperture glass filter
%
WAVELENGTH , nm
400nm
440nm
500nm
600nm
644nm
758nm
868nm
947nm
1024nm

Fig. 1

* A temperature sensor Si PTC (TP 1025), for recording the inside temperature of the instrument.

The sunphotometer is shown in Fig. 2. A front dial rotates the filter wheel in order to change the channels during the measurement. The aperture or optical window (8) for the incoming solar radiation is close to that dial. To reduce the influence of moisture and dust inside of the instrument the front aperture is protected with a suitable filter glas (Fig. 1). For aiming at the sun and getting an optical reading the instrument has a diopter (2) showing the position of the sun. An amplifier switch (6) with 5 ranges amplifies the detector current (7). The electronics for the amplification consist of a current to voltage converter and a voltage amplifier. The temperature stability of the used resistances is satisfactory. For the measurement of the residual solar intensity after the extinction in the atmopshere, the solar elevation angle, the inside temperature of the instrument and the power supply voltage, the instrument is provided with 3 push buttons "M", "U", "T" (4). It is necessary to record the inside temperature of the instrument, because the photodetector used becomes instable at high temperatures in its infrared range. An active temperature control could not be installed because of large power consumption.

A liquid crystal display (5) with 6 digits shows the measured data. The first three digits are reserved for the intensity measurement of the incoming radiation, while the last two sequentially indicate the solar elevation angle, the inside temperature of the instrument and the voltage of the rechargeable batteries. The display also shows during the measurements one decimal point whose position is used as indicator of the selected amplification. A plate (3) of 32 silicon solar cells transforms the sun radiation into electrical energy and supplies sufficient energy for the rechargeable batteries. In addition, the batteries can be recharged directly from commercial power. A switch (7) turns the instrument "ON" and "OFF". The instrument has furthermore some advantages to make the measurements easy and contribute to the accuracy of the measured values. For example, most photodiodes became wearied after an exposure time (VOLZ, 1961). To prevent from additional wear, the photodiode is protected from diffuse radiation between measurements with a shutter mounted on the filter wheel (1). This shutter can also be used to check for noise readings of the photodiode.

The opening aperture of the instrument is 57'. This way, most of the circumsolar radiation is shielded. Its contribution remains smaller than 0,3% at turbidity factors after LINKE of 6 (EIDEN, 1981) and 0,9% for optical thickness of 1 (LINKES turbidity factor 8,5) after GRASSL (1971), both resp. for 550 nm and 500 nm wavelengths. For a handheld instrument it is difficult to aim at the center of the sun for a reliable readout. The instrument thus is provided with an electronical memory for maximum values of incoming radiation, which is activated with one of the push buttons. Field measurements *) on a ship cruise in 1981 showed the instruments capability to perform measurements on a moving platform.

----

*) made by V. JOHANNES of our institute.

The different ranges enable the determination of turbidity values corresponding to visibility ranges of 100 m to more than 160 km.

The criteria for selection of the filters wavelengths depend on the nature of the informations which are wanted. In addition to the determinations of the aerosol optical depth, the instrument is able to measure the ozone optical depth from the broad Chappuis band in the visible region and the total amount of water vapor or precipitable water in the $\rho\sigma\tau$- water vapor band. Another trace gas in the visible spectrum is nitrogen dioxide which only occurs in polluted atmospheres. The amount of the attenuation of the sun radiation in the atmosphere is commonly described by the extinction which is defined by the Bouguer-Lambert law:

$$I(\lambda,h) = I_0(\lambda)\,\varepsilon\cdot \exp\left[-\tau_g(\lambda)\,M(h)\right] \qquad (1)$$

Related to the problem, one can write:

$$I(\lambda,h) = G\,\bar{T}_f(\lambda)\,\bar{T}_{gl}(\lambda)\,\bar{S}(\lambda)\,J_0(\lambda)\,\Delta\lambda\,\cdot \qquad (2)^*$$

$$\exp\left[-\left(\tau_R(\lambda)\frac{p}{p_0}+\tau_A(\lambda)+\tau_{gas}(\lambda)\right)M(h)\right]$$

where

$$G = (1-c)\;C\;A\;(\theta) \qquad (3)$$

and

$$\tau_{gas}(\lambda) = \tau_{O_3}(\lambda)+\tau_{H_2O}(\lambda)+\tau_{NO_2}(\lambda) \qquad (4)$$

(1) and (2) give:

$$I_0(\lambda) = G\;\bar{T}_f(\lambda)\,\bar{T}_{gl}(\lambda)\,\bar{S}(\lambda)\,J_0(\lambda)\,\Delta\lambda \qquad (5)$$

where

λ : wavelength, nm

$I(\lambda,h)$: spectral intensity of the direct solar radiation at ground level at wavelength λ

$I_0(\lambda)$: extraterrestrial intensity of the sun, which depends on the instrument as shown in eq. (5)

$\varepsilon = L^2/L_m^2$: relative distance between earth and sun

L_m : average distance between sun and earth

L : distance sun - earth at the observation time

---

*) scalar quantities with - such as $\bar{S}$ (λ) are averaged over the small bandwidth at $\Delta\lambda$ of the quasi monochromatic filters.

$M(h)$: relative air mass

h : sun elevation angle

$\tau_g(\lambda)$: total extinction coefficient

c : factor, which takes into account the electrical adjustment of the radiation receiver (photodiode), the A/D converter, the sensitivity of the amplifier resistance

C : conversion factor analog - digital, A^{-1}

$T_f(\lambda)$: spectral transmission of the interference filters

$T_{gl}(\lambda)$: spectral transmission of the aperture glas filter

$S(\lambda)$: absolute sensitivity of the detector, $A\ W^{-1}\ m^2$

$J_0(\lambda)$: extraterrestrial solar spectral irradiance, $W\ m^{-2} nm^{-1}$

$\Delta\lambda$: half bandwidth of the interference filter, nm

$\tau_R(\lambda)$: scattering coefficient of air molecule

$\tau_A(\lambda)$: extinction coefficient of dust particles

$\tau_{gas}(\lambda)$: extinction coefficient of atmospheric gas

p, p_0 : pressure values at the station and at NN

Equation (5) shows a relationship between the photometer constants, which don't take into account the absorption feature of the black tube. Some data like $S(\lambda)$ are large sources of errors. To obtain an absolute calibration of the photometer I_o is usually determined by a Langley plot $I_o(\lambda) = I(\lambda, M(h) = o)$. For this operation some atmospheric conditions are to meet: sky cover 1/8, good visibility during the whole constant turbidity day.

The measuring of $I(\lambda, h)$, h and the temperature of the inside of the instrument the following quantities can be determined:

* spectral turbidity factor after LINKE

* spectral extinction coefficient (optical depth)

* spectral decadic extinction coefficient, e.g. the turbidity coefficient after SCHÜEPP (SCHÜEPP, 1949)

* tubidity parameter after ANGSTRÖM (ANGSTRÖM, 1961)

* ozone optical depth (KING and BYRNE, 1976, SHAW, 1979)

* vertically averaged size distribution of atmospheric aerosol particles (YAMAMOTO and TANAKA, 1969, QUENZEL, 1970, BOX and SHUI YIN, 1979, HOYNINGEN-HUENE and SPÄNKUCH, 1979)

* quantitative determination of dust through a correlation analysis between turbidity and mass concentration of dust.

The last mentioned point is the more important goal of this investigation. For reaching this, the above described photometer is multiplicated and distributed at different stations in and southerly of the Sahara.

3. THE TURBIDITY NETWORK

Fig. 3 shows the stations of the turbidity network.
Tamanrasset and Assekrem (Algeria): The station of Tamanrasset which is 1380 m high is a station with available radiosounding data. This kind of data is needed for a study of the high atmosphere during weather perturbations such as sandstorm and dustwind and for the calibration of the water vapor channel used in the photometer. Assekrem (2730 m) is situated in the Hoggar-Mountains about 60 km NE from Tamanrasset. Both stations allow the determination of the latitude dependence of the aerosol and the atmospheric gases which is measured with the new photometer.

Boutilimit (Mauritania): The situation of this station is suitable for the determination of the frequency of sandstorm and dustwind and to quantify the aerosol total mass from the important potential sources such as Spanish Sahara, Southwest Algeria and North Mauritania. Together with weather maps and satellite images it would be possible to get an idea about the different aspects of the mineral dust transport from the Sahara towards the South and over the North Atlantic Ocean. The west component of the transport was investigated by several authors such as CARLSON and PROSPERO (1972), SCHÜTZ et al (1977), JAENICKE and SCHÜTZ (1978). The south component through Mauritania is minor and should be quantified.

Dakar (Senegal): This station is not suitable for a study of background aerosol because of its considerable anthropogenic sources. The upper atmosphere data which are available in this station is one of the reasons which support its choice. Furthermore there are most of the time different aerosol sources which influence the area of Boutilimit (Mauritania) and those of Dakar (Senegal) (d'ALMEIDA, 1979).

The line through Dakar (Senegal), Gao (Mali), Agades (Niger) and Wad Medani (Sudan) seizes all activities from the Sahara and the Sahel area. This enables the investigation of the Saharan Dust transport from the sources towards the South.

Korhogo (Ivory Coast), Tabligho close to Lomé (Togo) and Zaria (Nigeria) are situated more in the South. These 3 stations make possible the determination of the residual dust coming from the Sahara toward to the Golf of Guinea. It could be possible to get an idea about the dry and wet depostion in the region of the Intertropical Convergence Zone (ITCZ). Until now, it is not yet clear if and how the dust crosses the ITCZ. This study could give an answer. The area between Agades and Wad Medani is without stations. One or two stations in this important area would give a more complete picture. At the distribution

Fig. 2 The new sunphotometer with nine interference filters (described in section 2)

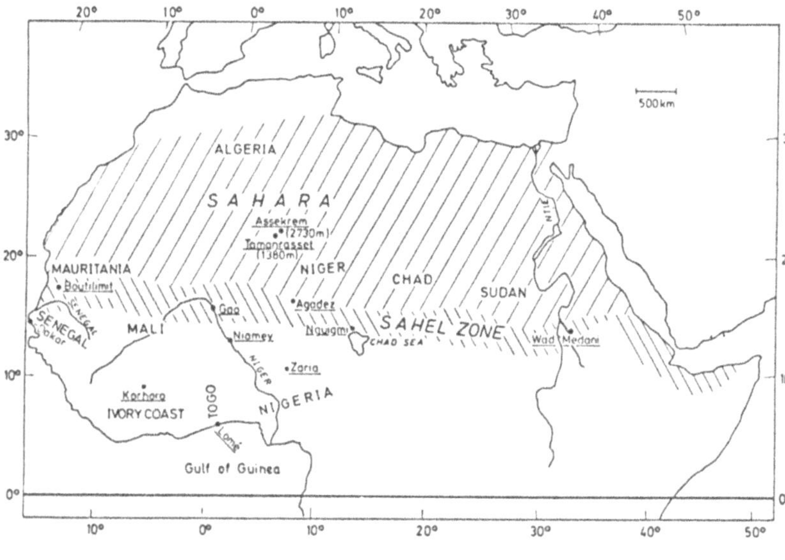

Fig. 3: STATIONS OF THE TURBIDITY NETWORK

time of the photometers there was an obscure political situation in Chad. A journey or an installation of an instrument in this country was unadvisable. In default of a station in Chad, it will shortly be possible to establish one photometer in N'guigui (Niger) near Lake Chad.

CONCLUSIONS

Sunphotometers still are a very versatile method of observing the atmospheric turbidity. This parameter is of importance for the local weather and global climatic considerations. In this special case, it can be used to determine the transport of Saharan dust to the South. The magnitude of this transport never was estimated, but might have implication with regard to the desertification of the Sahel region and the development of intertropical precipitation. It is estimated to have sufficient data after 2 years of operation of the network.

REFERENCES

D'ALMEIDA, G.A., 1979: Die vollständige mineralische Größenverteilung und Diskussion möglicher Quellgebiete von Sandstürmen der zentralen, westlichen und südwestlichen Sahara. D.A., Universität Mainz, FRG.

ANGSTRÖM, A., 1961: Technics of Determining the Turbidity of the Atmosphere. Tellus 13, No. 2, 214-223.

BOX, M.A., L. SHUI-YIN, 1976: Approximate Determination of Aerosol Size Distributions. J. Appl. Meteor. 15, 1068-1076.

CARLSON, T.N., P. WENDLING, 1977: Reflected Radiance Measured by NOAA & VHRR as a Function of Optical Depth for Sahara Dust. Appl. Meteor. 16, 1368-1371.

CARLSON, T.N., J.M. PROSPERO, 1972: The Large Scale Movement of Saharan Air Outbrakes over the North Equatorial Atlantic. J. Appl. Meteor. 11, 283-297.

EIDEN, R., 1981: Circumsolar Sky Radiation. 2nd. Solar Energy Programme of the Commision of the European Communities, Project F, Solar Radiation data (not published).

FOITZIG, L., H. HINZPETER, 1958: Sonnenstrahlung und Lufttrübung Akademische Verlagsgesellschaft Geest & Portig KG., Leipzig.

FRÖLICH, C., H. QUENZEL, 1974: Influence of the Sun's Aureole on the Determination of Turbidity. WMO-Publication 368, presented at TECOMAP, Helsinki.

GRASSL, H., 1971: Calculated Circumsolar Radiation as a Function of Aerosol Type, Field of View, Wavelength, and Optical Depth. Appl. Opt. 10, 2542-2546.

HOYNINGEN-HUENE, W., D. SPÄNKUCH, 1979: Zur Bestimmung von Aerosolgrößenverteilungen aus spektralen Extinktionsmessungen im Bereich von 340 nm bis 1100 nm Wellenlänge. Z. Meteor. 29, 147-156.

JAENICKE, R., L. SCHÜTZ, 1978: Comprehensive Study of Physical and Chemical Property of the Surface Aerosol in the Cap Verde Island Region. J. Geophys. Res. 83, 3585-3599.

KING, M.D., D.M. BYRNE, 1976: A Method for Inferring Total Ozone Content from the Spectral Variation of Total Optical Depth obtained with a Solar Radiometer. J. Atm. Sci. 33, 2242-2251.

KOEPKE, P., H. QUENZEL, 1978: Water Vapor: Spectral Transmission at Wavelengths between 0.7 µm and 1 µm. Appl. Opt. 17, 2114-2118.

KOEPKE, P., H. QUENZEL, 1979: Turbidity of the Atmosphere determined from Satellite Calculation of Optimum Viewing Geometry. J. Geophys. Res. 84, 7847-7856.

KOEPKE, P., H. QUENZEL, M. KAESTNER, 1980: Monitoring of Atmospheric Turbidity from Geostationary Satellites. Paper presented at the WMO Technical Conference on Regional and Global Observation of Atmospheric Pollution Relative to Climate. Boulder, Col. USA, 20 - 24 August, 1976, WMO No. 529.

MÖLLER, F., 193: Einführung in die Meteorologie, Bd. 2 B.I. Wissenschaftsverlag Mannheim/Wien/Zürich.

NORTEN, C.C., F.R. MOSHER, B. HINTON, W.D. MARTIN, D. SANTEK, W. KUHLOW, 1980: A Model for Calculating Desert Aerosol Turbidity over the Oceans from Geostationary Satellite Data. J. Appl. Meteor 19, 633-644

QUENZEL, H., 1966: Ein Interferenzfilter-Aktinograph zur optischen Bestimmung des atmosphärischen Gesamtwasserdampfgehaltes. Beitr. Phys. Atmosph. 39, 111-143.

QUENZEL, H., 1970: Determination of Size Distribution of Atmospheric Aerosol Particles from Spectral Solar Radiation Measurements. J. Geophys. Res. 73, 2915-2921.

SCHÜEPP, W., 1949: Die Bestimmung der Komponenten der Atmosphärischen Trübung aus Aktinometermessungen. Arch. Meteor. Geophys. Ser. B1, 257 - 346.

SCHÜTZ, L., 1977: Die Saharastaubkomponente über dem subtropischen Nord-Atlantik, Ph.D. Thesis University of Mainz, FRG.

SHAW, E.G., 1979: Aerosols at Mauna Loa: Optical Properties. J. Atm. Sci. 36, 862-869.

SHAW, E.G., 1979: Atmospheric Ozone: Determination by Chappuis-Band Absorption. J. Appl. Meteor. 18, 1325-1339.

VALKE, P. 1967: Über den Zusammenhang zwischen Trübungsfaktor und Trübungskoeffizient. Arch. Meteor. Geophys. Ser. B15, 357-375.

VOLZ, F. 1961: Photometer mit Selen-Photoelement zur spektralen Messung der Sonnenstrahlung und zur Bestimmung der Wellenlängenabhängigkeit der Dunsttrübung. Arch. Meteor. Geophys. Biokl. Ser. B10, 100-139.

YAMAMOTO, G., U. TANAKA, 1969: Determination of Aerosol Size Distribution from Spectral Attenuation Measurements. Appl.Opt. 8, 447-453

The Determination of the Atmospheric Heating Rate Due to the Absorption of Shortwave Radiation in Particles

Reinhold Busen, Gottfried Hänel, Christel Hillenbrand and Regine Schloß

Institut für Meteorologie und Geophysik der Johann Wolfgang Goethe-Universität, Frankfurt/Main, FRG

A b s t r a c t

To determine in situ the atmospheric heating rate, effected by absorption of solar radiation in aerosol particles, proceeding from a new theoretical derivation the shortwave radiation supply and the mean absorption coefficient of the aerosol particles have to be ascertained by measurements.
A radiometer was developed, which is used to measure the shortwave radiation supply, i.e. the shortwave radiation coming from all directions towards the point of measurement.
To measure the mean absorption coefficient of aerosol particles, averaged over the shortwave spectrum, two procedures are set up: Using a photometric equipment, particles collected on filters can be examined. A calorimetric technique is used for measurements on airborne particles. A third method, a calorimetric inspection of particles collected on filters, still is in the stage of development.
First measurements at different locations and with various weather conditions show substantial heating rates. Much higher heating rates are expected during the summer months, especially in heavy polluted urban and industrial regions.

1. INTRODUCTION

The global radiation budget in the multi-annual mean is fairly well known considering the top of the atmosphere, the surface of the earth and the total absorption in the atmosphere.

However, the amounts of shortwave radiation absorbed in the cloud free atmosphere and in clouds given by different authors (e.g. Paltridge and Platt (1976), Lockwood (1976)) move in the range of 8 to 20 percent of the incoming solar radiation for the first effect and 3 to 11 percent for the second one.

Because of these uncertainties in situ measurements of the absorption of shortwave radiation by gases and particles are desired. In the past beside numerous model calculations only a few measurements (e.g. Kondratyev(1976), Murai et al.(1976)) were made, but they were based on measuring the upwelling and downwelling radiation flux density in different altitudes and calculating the radiative heat flux divergence of the intermediate layers of the atmosphere. However, the heating rates derived in this way are correct only in a horizontal homogenous atmospheric layer of some extent, i.e. in a cloud free atmosphere and in altitudes about 200 m above the surface.

In the following new measuring techniques are described allowing in situ measurements of the amount of shortwave radiation absorbed by aerosol and cloud particles at every point of the atmosphere as well as at the ground.

2. THEORY

2.1 THEORETICAL DESCRIPTION OF THE ABSORBED AND TRANSFORMED RADIATION ENERGY

The energy $\dot{Q}_p$ of the shortwave radiation, which is absorbed in particles and transformed into heat, can be expressed as:

$$\rho \cdot c_p \cdot V \cdot \left(\frac{dT}{dt}\right)_s - \ell_{wp} \cdot \frac{dM_{wp}}{dt} = \dot{Q}_p \qquad (1)$$

ρ = air density

c_p = specific heat of air at constant pressure

V = volume of air

T = air temperature

t = time

ℓ_{wp} = heat of vaporization of water at constant pressure

M_{wp} = mass of water condensed on particles in the volume V

The first term describes the change in sensible heat, caused by radiation processes, the second one the change in latent heat due to the evaporation or condensation of water at the surface area of the particles.
Independent of this consideration the amount of energy $\dot{Q}_p$ can be obtained balancing the radiation processes at the surface area covering a small homogenous volume of air. The relevant formula reads:

$$\dot{Q}_p = - V \cdot (\mathrm{div}\, \vec{F})_p \tag{2}$$

where $\vec{F}$ is the net radiation flux density in the shortwave spectrum. Thereby the assumption is made, that the energy density of the shortwave electromagnetic radiation remains constant in the volume under consideration.
From the theory of radiation transfer $(\mathrm{div}\, \vec{F})_p$ can be expressed in the following way:

$$(\mathrm{div}\, \vec{F})_p = -\bar{\sigma}_{ap} \int_{4\pi\omega_1} N(\omega)\, d\omega \tag{3}$$

$\bar{\sigma}_{ap}$ = mean absorption coefficient of particles in the shortwave spectral region
N = radiance in the shortwave spectral region
ω_1 = unit solid angle

Combining (2) and (3) the shortwave energy per unit of volume and time, absorbed by aerosol particles and transformed into heat, can be expressed in the following manner:

$$\frac{\dot{Q}_p}{V} = \bar{\sigma}_{ap} \cdot \int_{4\pi\omega_1} N(\omega)\, d\omega \tag{4}$$

2.2 DISCUSSION OF THE NEW METHOD OF MEASUREMENT

As seen from equation (4), the mean absorption coefficient of aerosol particles $\bar{\sigma}_{ap}$ and the radiation supply, expressed by the integral term, have to be measured to determine the radiative energy being absorbed. The absorption coefficient is defined as the ratio of the absorbed to the incoming

radiation per cm. The radiation supply is the total shortwave radiation coming from all directions towards the point of measurement in W/m^2.
For illustration of the absorbed energy equation (4) can be written as:

$$\left(\frac{dT}{dt}\right)_p = \frac{\bar{\sigma}_{ap}}{\rho \cdot c_p} \cdot \int_{4\pi\omega_1} N(\omega)\, d\omega \qquad (5)$$

where $(dT/dt)_p$ is the atmospheric heating rate caused by the absorption of shortwave radiation in aerosol particles, expressed in K/d or better K/h. In this expression it is presupposed, that the liquid water content of the particles remains unchanged during the absorption of shortwave radiation.

3. MEASURING TECHNIQUES

3.1 SHORTWAVE RADIATION SUPPLY

The perfect measurement of the shortwave radiation supply would be to point a photometer successively to all directions and to integrate the readout of N. However, this method is technically too complicated and too time consuming, so we have developed a simplified measuring technique applicable for routine measurements and continuous recordings. With the

Figure 1: Schematic diagram of the integrating radiometer

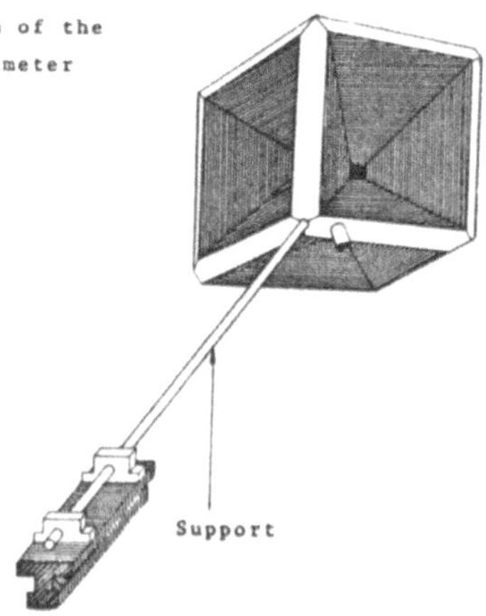

new instrument, an integrating radiometer, the flux densities of shortwave radiation coming from six equal ranges of the solid angle are measured separately by thermopiles covered with a quartz disk (Fig.1). The sum of these six ranges of the solid angle is equal to 4π sr.

The flux densities measured are corrected

a. for the error (about 3.5%) arising from the fact that radiation flowing from an edge of a funnel to one of the sensors also can be determined partly by the sensor detecting the adjacent range of the solid angle, and
b. for the error (about 25 ±5%) coming from the cosine response of the sensors.

To avoid the large cosine error for the direct solar radiation the solar radiant flux is measured with an actinometer by Linke and Feussner, mounted on a solar tracker. With the actinometer readout the output of the integrating radiometer is corrected.

The total error by which the shortwave radiation supply can be measured is smaller than ±7.2% during sunshine and smaller than ±8.6% when there is no sunshine. These errors have been computed considering the calibration errors of the instruments and the distribution of the shortwave sky radiation and the reflected shortwave radiation from walls and from the ground according to the literature and own calculations. At present, the integrating radiometer is modified slightly to reduce considerably the cosine error.

3.2 ABSORPTION MEASUREMENTS

3.2.1 PHOTOMETRIC METHOD FOR SAMPLES OF PARTICLES ON FILTERS

For the photometric absorption measurement particles from the atmosphere are collected on a membrane filter. In the apparatus the particle loaded filter is placed in front of a light diffuser and the particles are directly irradiated with shortwave radiation from a lamp emitting radiation close to the solar spectrum (Fig.2). The flux density of the

Figure 2: Schematic diagram of the photometric absorption measurement

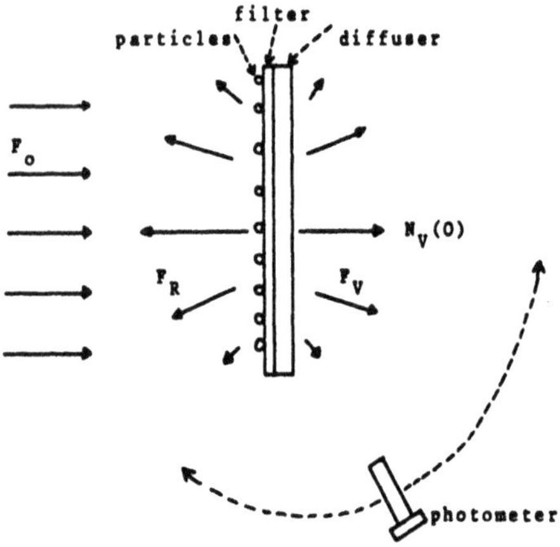

incoming radiation and the angular dependence of the radiation scattered by the system particles-filter-light diffuser is measured with a thermopile covered with a quartz plate. For comparison purposes this measurement is repeated for the system filter-light diffuser alone.
Using energy budget equations for both systems and considering multiple reflections of radiation between the particles and the system filter-light diffuser we get that part of the flux density of the incoming parallel radiation being absorbed by the particles.
According to Chin I Lin et al. (1973) the absorption of radiation by the particles on the filter can be equalized to the absorption of radiation by the same particles in airborne state, whennot the particle loading on the filter is too high. So the absorption coefficient can be expressed as

$$\bar{\sigma}_{ap} = -\frac{A}{V} \cdot \ln (1-F_{ap}/F_o) \qquad (6)$$

A = particle loaded area of the filter
V = volume of air sucked through the filter
F_{ap} = flux density of radiation absorbed by particles
F_o = flux density of the incoming radiation

The mean absorption coefficient obtained has to be corrected, because the wavelength dependencies of the shortwave radiation emitted by the lamp and of the shortwave radiation supply under real atmospheric conditions are slightly different. It is

$$\bar{\sigma}_{ap}(\text{atmosphere}) = (0.955 \pm 0.010) \cdot \bar{\sigma}_{ap}(\text{lamp}) \qquad (7)$$

With the photometric method also the extinction coefficient and the ratio of forward and backscattering of the particles have been measured. Thus with this method all the aerosol properties needed for radiative transfer calculations with two stream approximations are obtained.

3.2.2 CALORIMETRIC METHOD FOR AIRBORNE PARTICLES

The experimental setup for the calorimetric method mainly consists of two closed chambers being connected by a differential pressure meter of high sensitivity (Fig.3). Into one of these chambers (the measuring chamber) an absorber plus dry air are sucked, into the second one (the comparison chamber) dry air only. When these chambers are illuminated at the same time by parallel shortwave radiation, electromagnetic energy is transformed into internal energy within the

<u>Figure 3</u>: Schematic diagram of the calorimetric absorption measurement using airborne particles

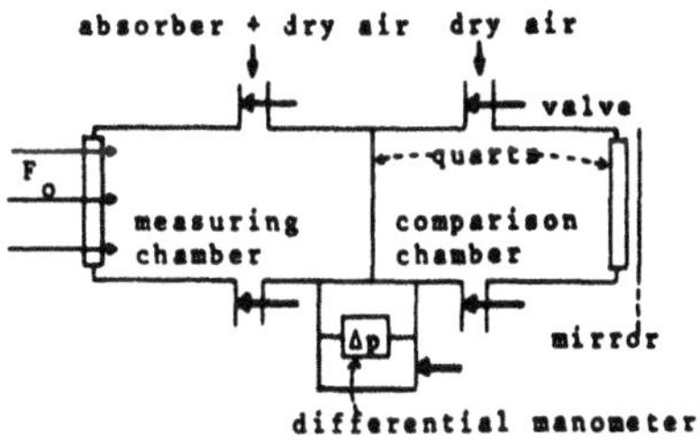

absorber. When this energy is transferred to the surrounding dry air a pressure increase in the measuring chamber takes place, which is directly proportional to the flux density absorbed. The relevant formula reads

$$1 - \frac{F_{ap}}{F_o} = 1 - \frac{c_v \cdot V \cdot \Delta p_{ap}}{R \cdot F_o} = \exp(-\bar{\sigma}_{ap} \cdot \ell) \qquad (8)$$

where $V = 524\ cm^3$ = volume and $\ell = 18$ cm = length of the measuring chamber, c_v = specific heat of the air at constant volume, R = specific gas constant of the air, Δp_{ap} = pressure increase in the measuring chamber due to the absorption of shortwave radiation within the particles.

3.2.3 CALORIMETRIC METHOD FOR SAMPLES OF PARTICLES ON FILTERS

Clean air sucked through a clean filter will be heated, if the filter is illuminated by shortwave radiation and therefore is warmer than the surrounding air due to the absorption of the shortwave radiation. Being loaded with aerosol particles the filter will absorb more radiation energy than in the unloaded stage and so the temperature increase of the air passing the filter in the first case will be higher than in the last one. The experimental setup is shown in Fig.4. The approximate absorption coefficient is obtained by

$$\bar{\sigma}_{ap} \cong \frac{c_p}{F_o \cdot V_1} \cdot (\dot{M}_p \cdot \Delta T_p - \dot{M} \cdot \Delta T) \qquad (9)$$

where F_o = flux density of the shortwave radiation at the filter, V_1 = volume of air sucked through the filter when sampling the particles, $\dot{M}$ and $\dot{M}_p$ = mass flow rate of the air in the case of the unloaded and loaded filter and ΔT and ΔT_p = temperature increase of the air in both cases.
Up to now measurements of the shortwave absorption coefficient not have been made in this manner, because this method still is in the stage of development.

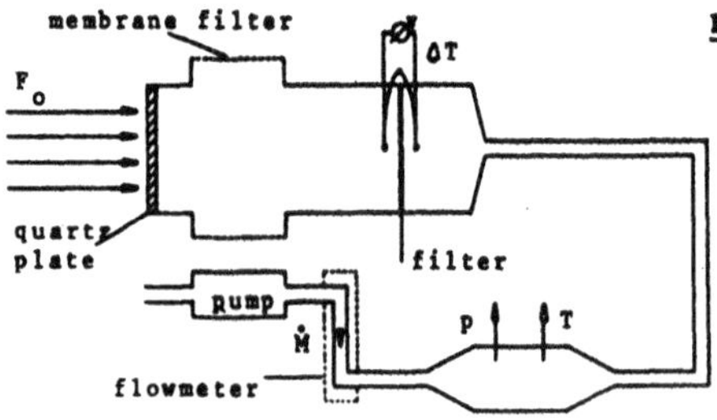

Figure 4: Schematic diagram of the calorimetric absorption measurement using particles on filters

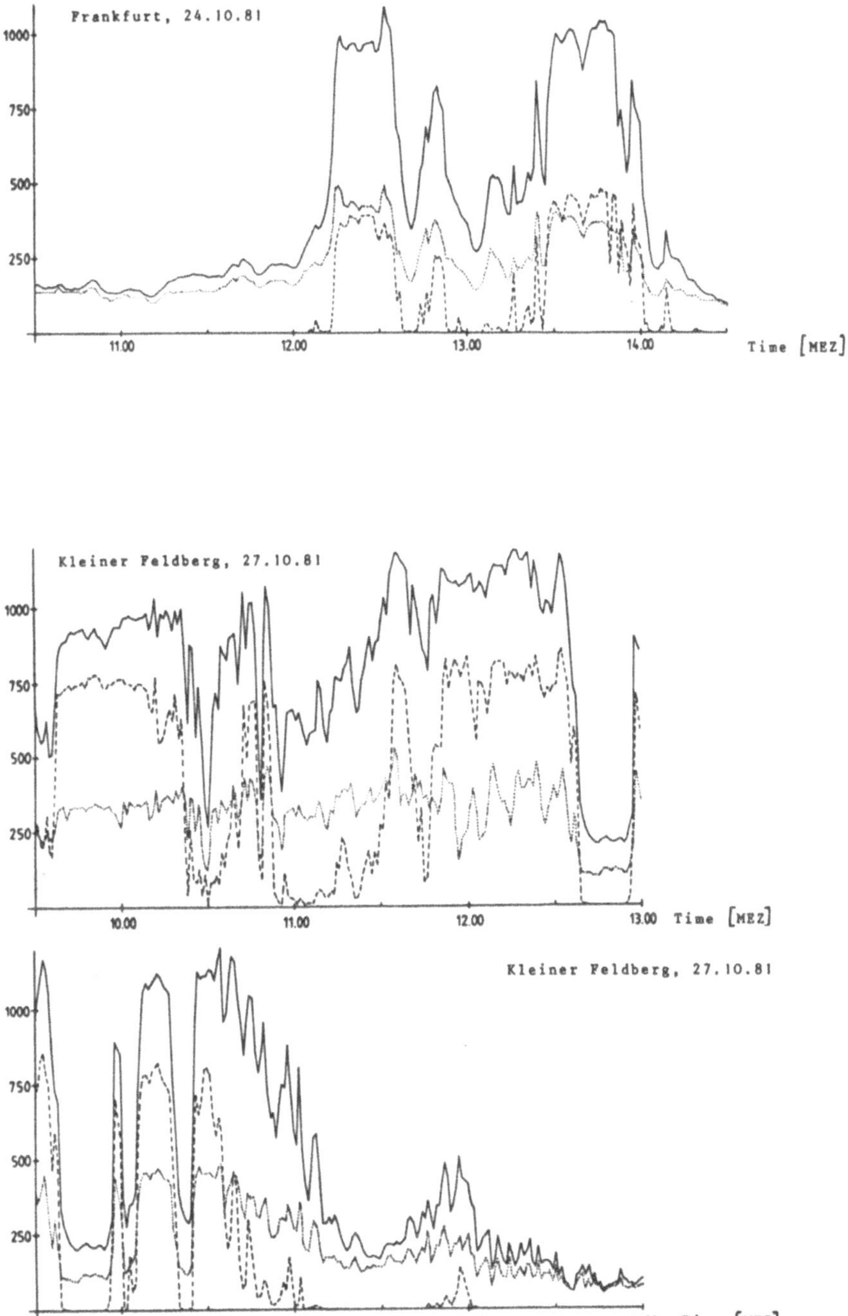

Figure 5: Shortwave radiation supply $\int_{4\pi\omega_1} N(\omega)\, d\omega$ in W/m^2 (solid line)
Dashed line: Flux density of direct solar radiation
Dotted line: Shortwave global radiation

4. RESULTS OF MEASUREMENTS

In Fig.5 some examples of measurements of the shortwave radiation supply are compiled. For comparison also the flux density of the direct solar radiation and the global radiation, measured with a pyranometer made by Kipp&Zonen, are plotted. The variations of the shortwave radiation supply have to be interpreted as real variations of the radiation field appropriate to the excellent correlation with the global radiation. The variations are caused by currently changing cloud formations on 24.10.81 at Frankfurt, whereas on the top of Kleiner Feldberg/Taunus veils of mist were the reason.

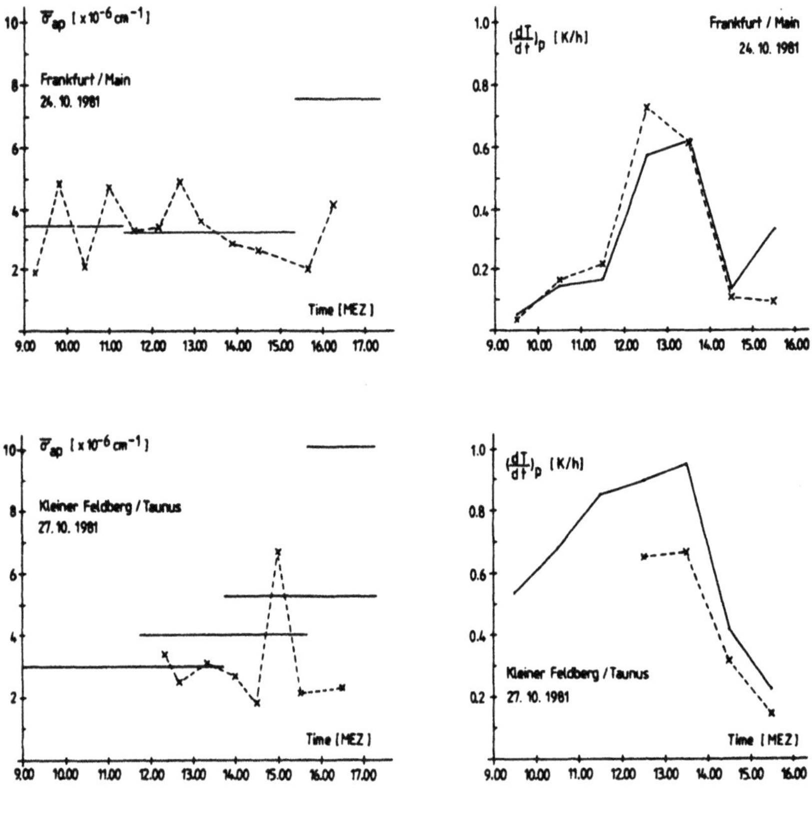

Figure 6 : Mean shortwave absorption coefficients $\bar{\sigma}_{ap}$ and mean hourly heating rates.
Measurements and calculations based on the photometric method (solid lines) and on the calorimetric method (dashed lines).

Mean absorption coefficients, measured on both days with the photometric equipment (solid lines) and with the calorimetric method (crosses, connected by dashed lines), are shown in Fig.6. Additional the mean hourly heating rates, calculated by combining the radiation supply and the absorption coefficients, are plotted. A real good agreement is found comparing the results following from the two fundamental different and independent absorption measurements. The heating rates increase up to 1.0 K/h and reach their highest values at noon corresponding to the elevation of the sun. The total heating, refered to the daytime period and to the photometric absorption measurements, amounts to 2.03 K on 24.10. and to 4.55 K on 27.10.

5. DISCUSSION

At present the instrumentation described here is in further development. The radiometer is modified to reduce the cosine error and to diminish the weight of the instrument for vertical soundings of the atmosphere using a balloon equipment. Measuring the mean absorption coefficient of aerosol particles in three completely different ways, it is possible to check the methods reciprocal.

Already the few results presented here show clearly the high climatological importance of the absorption of solar radiation in aerosol particles. Regarding Rodgers (1972) and Zdunkowski and McQuage (1972) these heating rates have substancial effects on weather forecast as well as on the vertical profiles of wind and temperature in the boundary layer. Since the most absorbing particles have sizes in the order of 0.1 μm and thus can be transported over very long distances, the atmospheric heating due to the absorption of shortwave radiation in particles cannot be valued as a local phenomenon alone.

Proceeding from the results of measurements and assuming clear sky conditions, even during the winter time with low sun elevations the heating rates can easily exceed 1.0 K/h

During the summer months they are expected to exceed 2.0 K/h and more, especially in heavy polluted urban and industrial regions. Then the daily total heating should reach 10 - 15 K because of the higher sun elevations and the extended period of daylight.

References

Chin-I Lin, Baker, M., Charlson, R.J. (1973):
Absorption Coefficient of Atmospheric Aerosol: A Method for Measurement. Applied Optics 12, No.6, 1356 - 1363

Hänel, G. et al. (1982):
Light absorption measurements: New techniques. Applied Optics 21, 382 - 386

Kondratyev, K.Y. et al. (1976):
Aerosol in the GATE Area and Its Radiative Properties. Atmospheric Science Paper No. 247, Colorado State University, Fort Collins

Lockwood, J.G. (1976):
Causes of Climate. Edwald Arnold Ltd., London

Murai, K. et al. (1976):
The absorption of solar radiation in the lower atmosphere Papers in Met. and Geoph. 27, 21 - 32

Paltridge, G.W. and Platt, C.M.R. (1976):
Radiative Processes in Meteorology and Climatology. Elsevier Scientific Publishing Company, Amsterdam - Oxford - New York

Rodgers, C.D. (1972):
Parameterization of sub-grid scale processes. GARP Publications Series, WMO, No.8, 6 - 19

Zdunkowski, W.G. and McQuage, N.D. (1972):
Short-term effects of aerosol on the layer near the ground in a cloudness atmosphere. Tellus 24, 237 - 254

New Meteorological Applications of the Aerosol Physical Chemistry

Gottfried Hänel

Institut für Meteorologie und Geophysik der Johann Wolfgang Goethe-Universität, Frankfurt/Main, FRG

1. INTRODUCTION

During the last years our knowledge about the physico-chemical properties of atmospheric aerosol particles has increased so much (Hänel 1976, Hänel and Lehmann 1981) that at present enough information is available allowing reliable meteorological applications. In this paper two processes, the settlement of particles through the atmosphere under the influence of gravity and the activation of particles in fog and clouds, are considered.

2. MEAN SETTLING VELOCITY OF PARTICLES AND RELATIVE HUMIDITY

There are two major sinks of atmospheric particles, the coagulation, reducing mainly the concentration of the smaller particles by agglomeration due to Brownian diffusion, and the settlement of particles through the moving air because of the action of gravity, diminishing mainly the concentration of the largest particles. In the past much work has been done to understand coagulation. However, the settlement has not been understood as well. It is the aim of this paper to study the influence of relative humidity on the settlement of particles in the atmosphere.
For the three main aerosol types in the atmosphere, maritime, urban, and background particles the mean settling velocity for dry material

$$\bar{w}_s = \sum_{j=1}^{1} \bar{m}_j \bar{w}_{sj} \Bigg/ \sum_{j=1}^{1} \bar{m}_j = \frac{\text{mass flux density}}{\text{mass concentration}} \qquad (1)$$

$\bar{w}_{sj}$ = amount of the mean relative velocity between particles of size class j and the moving air in the vertical direction = settling velocity of particles in stagnant air after a residence time of the particles in the atmosphere between 2 and 30 sec.

$\bar{m}_j$ = mean mass concentration of dry material, lead, iron, cadmium or manganese of size class j in air(in g cm^{-3}).

has been computed as a function of relative humidity. Additional computations have been performed for lead, iron, cadmium, and manganese in an urban atmosphere. The values of $\bar{w}_{sj}$ have been computed following the procedure of Hänel 1977, however using the drag coefficients by Pruppacher and Steinberger 1968 and dynamic shape factors as functions of relative humidity after Hänel 1977, Pettyjohn and Christiansen 1948, and Stöber et al. 1970.
The experimental physico-chemical aerosol properties have been taken from Hänel 1976 and the experimental concentration distribution of the metals at Frankfurt from Müller 1980. The results are compiled in Tables 1 and 2.

<u>Table 1:</u> Mean settling velocity $\bar{w}_s$ of dry material as function of relative humidity f.

a)

Aerosol type	$\bar{w}_s$ (f = 0) (cm / sec)
Maritime	.214
Urban	1.07
Background	.785

b)

f (per cent)	$\bar{w}_s(f)$ / $\bar{w}_s(f = 0)$ Maritime	Urban	Background
40	1.009	1.007	1.011
80	1.60	1.30	1.34
95	2.62	2.05	1.93
99	5.98	4.81	4.44
99.9	24.8	16.5	16.6

Table 2: Mean settling velocity $\bar{w}_s$ of metals at Frankfurt as function of relative humidity f.

f (per cent)	$\bar{w}_s$ (cm / sec)			
	Lead	Iron	Cadmium	Manganese
0	.31	1.6	.17	.76
40	.31	1.6	.17	.76
80	.40	2.1	.22	.98
95	.64	3.3	.34	1.6
99	1.5	7.8	.80	3.7
99.9	5.3	27	2.8	13

These results allow the following conclusions:

a) There is a gradual increase of the mean settling velocity with relative humidity. This increase becomes very steep when the relative humidity is larger than about 90 per cent. Thus in this region there is a gradual change-over from 'dry' to 'wet' removal of particles from the atmosphere.

b) Episodes of high relative humidity during nighttime and of fog (99.9 per cent relative humidity) are very effective in removing particulate matter from the atmosphere. During 1 to 1.5 hours of fog the same quantity of material can be deposited on the ground by settlement as during a whole day of small or moderate relative humidity.

Consequently for a reliable modelling of the deposition of particulate matter on the ground the effect of relative humidity has to be considered.

3. INFLUENCE OF PHYSICO-CHEMICAL PROPERTIES OF ATMOSPHERIC AEROSOL PARTICLES ON THEIR ACTIVATION IN FOG AND CLOUD

When a parcel of atmospheric air is cooling - either by ascending in the atmosphere or by radiative effects - the relative humidity increases. Then water vapor condenses on the particles and they grow. In fog and cloud this condensation process limits the maximum relative humidity to values being only slightly larger than 1 (i.e. 100%). Because of this only a fraction of particles - the so-called activated particles - can grow continuously by condensation. The remaining particles - the nonactivated particles - are practically in thermodynamic equilibrium with the surrounding moist air. Since the activated particles govern the precipitation formation, their number concentration must be known. Secondly the number concentrations of the activated and nonactivated particles determine the optical properties of fog and cloud and consequently the radiative transfer therein. In the past Fitzgerald 1974 and In Lee, Hänel, and Pruppacher 1980 have undertaken preliminary studies of the influence of chemical composition and size distribution of particles on the maximum relative humidity, the number of activated particles, and the liquid water content of a cloud. The results show a large influence of the aerosol properties on the number of activated particles and maximum relative humidity. However, due to the immense effort calculating non-equlilibrium growth of particles to cloud droplets no systematic investigations could be undertaken. Therefore a simple parameterization has been derived to compute the influence of the aerosol properties on the number of activated particles more systematically (Hänel 1981). The physical basis of this parameterization is that the non-activated particles are assumed to be in thermodynamical equilibrium with the surrounding moist air and that there are only small changes of the aerosol properties.

Since the details of the parameterization and the results have been published already, here only a summary of the meteorological consequences shall be given. At first, the most important results for constant maximum relative humidity are summarized:

a) The number of activated particles increases with temperature. This increase is largest for urban particles (about 18 per cent / 10°C).

For maritime and background particles the variation of the number of activated particles with temperature should be negligible in the most cases, however, for very small supersaturation it also can attain values up to about

18 per cent / $10^{o}C$. Consequently the temperature should have the largest effect on the formation of clouds in urban areas.

b) The effect of the chemical composition on the number of activated particles is largest in polluted areas. There the number of activated particles can vary by 300 to 500 per cent at least. For specific maritime and background aerosols the number of activated particles can be practically independent from the chemical composition of the particles.

Secondly the sensitivity of the number of activated particles on a small change - an increase by one per cent - of one of the aerosol parameters has been studied. Now the aerosol parameters influence the maximimum relative humidity. When the increase of an aerosol parameter leads at constant maximum relative humidity to an increase (a decrease) of the number of activated particles this model inplicates simultaneous increases (decreases) of the liquid water content and the temperature of the cloud as in the atmosphere. These decrease (increase) the maximum relative humidity and indirectly also the number of activated particles. Now the increase of an aerosol parameter leading at constant maximum relative humidity to an increase of the number of activated particles can result even in a diminution of the number of activated particles depending on the meteorological conditions. The results of this model allow some additional meteorological conclusions:

a) Variations of the aerosol parameters result in the largest changes of the number of activated particles when the meteorological conditions yield small maximum supersaturations or large liquid water contents. Thus the properties of fog, stratus clouds, and very active cumulus clouds should be highly dependent on the physico-chemical properties of the aerosol particles.

b) Depending on the meteorological conditions the number of activated particles may increase or decrease due to a change of only one aerosol parameter.

During cloud seeding many aerosol properties are changed at the same time. This explains the difficulties predicting its consequences.

c) Both the physico-chemical and the size distribution parameters are of significance. The relative importance of one parameter compared to the others depends on the initial values of the inherent properties of the atmosphere and of the aerosol particles.

Summarizing, the results of this study show that realistic modelling of fog, cloud, and precipitation formation - consequently also modelling of the effects of cloud seeding - seems to be impossible without consideration of the microphysical parameters of the aerosol particles. Therefore, the properties of fog and cloud - inclusive their radiative properties - and thus the role of fog and cloud in climate cannot be regarded to depend only on the macrophysical meteorological conditions prior to fog and cloud formation.

4. REFERENCES

FITZGERALD, J.W., 1974: Effect of Aerosol Composition on Droplet Size Distribution: A Numerical Study, J. Atm. Sci., 31, 1358-1367

GEORGII, H.-W., et al., 1980: Untersuchungen über die trockene und feuchte Deposition von Luftverunreinigungen in der Bundesrepublik Deutschland. Forschungsbericht des Instituts für Meteorologie und Geophysik, Frankfurt/Main

HÄNEL, G., 1976: The Properties of Atmospheric Aerosol Particles as Function of the Relative Humidity at Thermodynamic Equilibrium with the Surrounding Moist Air, Advances in Geophysics, 19, 73-188

----, 1977: Humidity Effects on Gravitational Settling and Brownian Diffusion of Atmospheric Aerosol Particles, PAGEOPH 115, 775-797

----, 1981: Influences of Physical and Chemical Properties of Atmospheric Particles on the Activation Process in Fog and Clouds, Beitr. Phys. Atmos., 54, 159-172

----, and M. LEHMANN, 1981: Equilibrium Size of Aerosol Particles and Relative Humidity: New Experimental Data from Various Aerosol Types and Their Treatment for Cloud Physics Application, Beitr. Phys. Atmosph., 54, 57-71

IN LEE, Y., G. HÄNEL, and H.R. PRUPPACHER, 1980: A Numerical Determination of the Evolution of Cloud Drop Spectra due to Condensation on Natural Aerosol Particles, J. Aerosol Sci., 11, 251-152

PETTYJOHN, E.S., and CHRISTIANSEN, E.B., 1948: Effect of Particle Shape on Free Settling Rates of Isometric Particles. Chem. Engng. Prog. 44. 157-172.

STÖBER, W., et al.,1970: Der aerodynamische Durchmesser von Latexaggregaten und Asbestfasern. Staub 30, 277-285.

On the Flux and Collision Mechanism of the Scavenging Process of Atmospheric Aerosol Particles

F. Herbert

Institute of Meteorology and Geophysics, J. W. Goethe-University Frankfurt/Main, FRG

S u m m a r y : The motion and transport of aerosol particles as well as the collection mechanism between water drops and particles in an air-vapor atmosphere are subjected to a series of dynamical driving forces. A practicable mathematical treatment is developed by applying the method of irreversible processes in connection with specialized momentum laws for the atmosphere and the particles and useful expressions for various transport parameters. Collection kernels for aerosol particles with radii $0.001 \leq a \leq 10$ µm colliding with singular water drops with radii $10 \leq a \leq 500$ µm are explicitly computed and discussed with respect to spherical symmetry.

1. RECENT MODEL ATTEMPTS

Meteorological content of this paper is a description of the physical mechanism leading to the removal of natural and anthropogenic aerosol particles from the atmosphere due to coagulation interactions with cloud and small precipitation drops. Observations and practical experiences show that this removal process is affected by a number of driving forces resulting in a complex flux and collision mechanism. The literature of the past two decades up to about the early 1970's provided a series of substantial studies to the understanding of the atmospheric aerosol scavenging mechanism; however, these previous models include considerable deficiencies particularly in the following aspects:

1) they neglect altogether or describe quite unexactly hydrodynamic and inertia effects which especially play a role in the caption of aerosol particles by falling water drops,
2) they ignore the eminent influence to the particle collection by Coulomb and Joule effects which act if the aerosol particles and the collector drops are electrically charged as well as imbedded in an external electric field,
3) they partially neglect the inclusion of thermo- and diffusiophoretic driving forces on the particle motion which are effective within a drop-condensing or drop-evaporating air-vapor mixture,
4) they compute the total scavenging efficiency on the assumption that the collection kernels of the individual collision and capturing processes are additive rather than on the basis of the actual transport and impaction mechanism which considers the combined, simultaneous action of all scavenging effects.

Various attempts to eliminate the physical deficiencies in previous models have been performed in more recent papers by Beard and Grover (1974), Hall and Beard (1975) and Grover (1976 a,b) who undertook considerable efforts to solve the aerosol scavenging problem with the aid of a systematical treatment of the aerosol motion and collision (collection) mechanism.

In related studies these investigations were used by Grover et al. (1977) and Wang et al. (1978) to formulate two complementary mathematical models for computing the collection efficiency with which relatively large and relatively small aerosol particles are captured by collision with water drops in the air, due to the simultaneous action of various driving forces. These two models which improve on some of the earlier approximate descriptions of particle scavenging by water drops in the air are based on quite different physical concepts.

The first model presented by Grover et al.(1977) uses the so-called 'trajectory method' adopted for relatively large aerosol particles with radii greater than about 0.5 µm. It is applied to quantitative computing the collection efficiency between spherical aerosol particles and spherical cloud droplets and small raindrops due to the combined, simultaneous action of inertial impaction and phoretic and electric driving forces. The various physical effects are joined together in a hypothetical equation of motion for an electrically charged and evaporating water drop in air. In this equation of motion the terms by gravitational attraction, hydrodynamic drag and electric and phoretic forces are balanced by an inertia force term due to the particles acceleration. The 'trajectory model', however, does not include the collection efficiency of aerosol particles by Brownian diffusion. Actually, the trajectory method thus applies to a single nondiffusing Newtonian particle whose velocity, explicitly time dependent, is determined on its path around the collector and scavenging drop. This 'single particle method' completely neglects thermo-kinetic driving forces as a result of Bownian motions which arise if the aerosol particles are considered as a part of a statistical ensemble which permanently interacts with the surrounding air; nevertheless, diffusio - and thermophoretic forces with quite similar non-Newtonian character like Brownian diffusion are included into the theoretical approach. This discrepancy between a Newtonian single particle and a half-thermodynamic ensemble formulation without Brownian diffusion forcing makes the trajectory model not particularly reliable.

In contrast to a quasi-Newtonian 'single-particle equation of motion' used in the trajectory model, the second model is based on a flux-transport equation for an ensemble of Brownian particles which are influenced by phoretic and Coulomb driving forces. This so-called flux model, applied by Wang et al. (1978), intends to supplement the trajectory model with the aim to allow the quantitative computation of the collection kernel or collection efficiency between spherical aerosol particles and spherical water drops in air due to the combined, simultaneous action of Brownian diffusion and phoretic and Coulomb forces. Since this model does not account explicity for inertia impaction effects by particle acceleration and gravity forces it is applied by the authors to the scavenging process of relatively small aerosol particles with radii smaller than about 0.5 µm. With the inclusion of Brownian and phoretic forces the flux model considers all stochastic impaction forces which describe the stationary mass transport of a particle ensemble within the surrounding air-water vapor mixture. These purely thermodynamic effects are caused by the system's particles, vapor and temperature inhomogeneities. In contrast, Coulomb force impaction due to the presence of electric charges on the aerosol particles and the singular collector drops has a quite different physical origin and can not follow from commonly thermodynamic considerations. Indeed, the presence of external field forces is due to the thermoelectromagnetic state of the air-vapor-particles system. However, Coulomb forces between charged aerosol particles and waterdrops are dynamically connected with the momentum law of the single particles and thus have the meaning of an indirect driving force which is primarily part of the equation of motion of one particle. This behavior is quite similar to that of gravitational sedimentation and buoyancy forces which are secondary effects becoming indirectly present in diffusion flux equations by substituting the original pressure forces for the case of mechanical equilibrium (Herbert, 1980). With this reasoning one may conclude that Wang et el.'s flux model really includes some type of inertial impaction force; therefore this model probably applies in principle, at least with respect to its flux-force approach, both to small and large aerosol particles.

In order to decide whether such a conclusion is justified or not the total particle flux and collection mechanism under the complex atmospheric influence must be reexamined in terms of a more stringent and closed theoretical study than those of Wang et al. and Grover et al.. From the comparison of these recent scavenging models follows that in order to arrive at a theoretical approach general enough to apply to both particle ranges to which the trajectory and flux model apply, it is necessary to reformulate the trajectory method by adding physical concepts which are explicitly taylored to an ensemble of aerosol particles. It seems mandatory to reformulate the flux method with the help of a mass transport theory which explicitly considers the kinetic energy of aerosol particles due to their own motion in addition to the other relevant energy terms. One avenue along which one derives a general model equally verifying the particles motion and transport in the trajectory and flux method has been discussed by Herbert (1978). Therein, particle dynamics is adopted to obey a somewhat modified and extended 3-component mixture consisting of dry air, water vapor and aerosol particles each of which, however, moves as a Brownian particle with its own kinetic energy. In particular,the study considers the fundamental criterion that the total entropy of the air-vapor-particle system increases with time due to dynamical and thermal irreversible interactions within the system's interior. Compared with earlier, quite hypothetical attempts which merely assume a dynamical concept for particles from portions of the total size range, this treatment has the considerable advantage to hold for aerosol particles of the whole size spectrum. This allows to predict a general equation for the particles dynamical behavior which equally applies to the small and large size range, i.e. to particles with radii between about of 0.oo1 up to < 100 µm. As it follows from Herbert (1978) this theoretical analysis provides a mathematical relation which explicitly connects all driving forces acting on the particles and the total particles current density

$$\vec{I}_p \equiv \rho x_p \, (\vec{v}_p - \vec{u}) \quad ,$$

where x_p and $\vec{v}_p$ are the mass fraction and the mean barycentric velocity of the aerosol particles while ρ and $\vec{u}$ represent the density and the barycentric velocity of the total system.

For $\vec{I}_p$ one may use the customary relation

$$\vec{I}_p = - \rho D_p \nabla x_p + \rho x_p B_p \sum_k \vec{F}_k \tag{1}$$

where D_p is the diffusion and B_p the mobility coefficient of the particles while $\Sigma \vec{F}_k$ represents the sum of all non-Brownian or so-called external driving forces acting on the particle flux (Herbert,1978). It then follows that

$$\vec{I}_p = - \rho D_p \nabla x_p + \rho x_p B_p \left\{ -m \left[(v_p - v_a) \nabla p - \vec{F}_e + \frac{d(\vec{v}_p - \vec{u})}{dt} \right] + \vec{F}_v + \vec{F}_t \right\} \tag{2}$$

where all 'external' forces $\vec{F}_k$ are specified in usual notation. One finds that two essential features of the particles dynamical behavior are explicitly expressed by the structure of equation (2). It is shown that
1) the total 'external' driving force of the particle current density consists of the pressure forces and inertia forces due to self-motion of the particles ($\vec{v}_p$) and the motion of the system ($\vec{u}$) as well as of electric field forces ($\vec{F}_e$) and diffusio- and thermophoretic forces ($\vec{F}_v$ and $\vec{F}_t$), and
2) that conversely, the particles current density $\vec{I}_p$ in its general formulation does not explicitly depend on gravity and Coulomb forces, a result which verifies a corresponding plausibility hypothesis argued in the discussion above.

2. A THEORETICAL (APPROXIMATE) SOLUTION

For the following investigations Equation (2) is equivalently evolved in a real flux equation for the particle current density $\vec{I}_p$. This procedure becomes eminently plausible when the dependent inertia force terms in (2) are substituted in terms of independent, quasi stationary forces acting on the particles.
Actually, both inertia forces are associated with mechanical momentum laws which may be defined under various assumptions on the behavior of the atmosphere. In any case, however, it might be realistic to assume that the barycentric motion $\vec{u}$ of the atmosphere depends on the combined action of pressure forces (∇p), friction forces ($\vec{F}_f$) and gravity and electric field forces ($\vec{g}$ and $\vec{F}_e$) while Coriolis forces might be omitted under cloud physical conditions. In particular, this study considers the special case that the motion of aerosol particles within air toward a collector drop merely causes a negligibly small impact on the mass or pressure into a basic state and perturbation pressure is therefore not required. For air moving as a result of these external and internal conditions one may formulate the following equation of motion

$$\frac{d\vec{u}}{dt} = - \frac{\nabla p}{\rho_a} + \vec{F}_f + \vec{g} + \frac{\rho_p}{\rho_a} \vec{F}_e \quad . \tag{3}$$

The quantities ρ_p and ρ_a are the partial densities of the aerosol particles and air where $\rho_p \ll \rho_a$. In addition to the barycentric momentum of air we also must describe in (2) the momentum of a single aerosol particle. For this purpose we hypothesize that the momentum of a particle of mass m with velocity $\vec{v}_p$ and a certain electrical charge obeys a quasi-Newtonian law of the form

$$\frac{d\vec{v}_p}{dt} = \vec{g} + \vec{F}_e + \frac{\vec{F}_c}{m} \quad (4)$$

which considers external forces by the earth's gravity and electric and magnetic fields as well as Coulomb forces, $\vec{F}_c$, occuring due to the singular action on particles moving toward a charged collector drop. Note that such Coulomb forces are not caused by an external field and thus may not be assumed to provide a responsible force in the equation of motion of the atmosphere (3).

To round out the atmospheric adjustment of the mechanical effects in the particle flux equation (2) we finally use the approximation

$$(v_p - v_a)\ \nabla p \approx - \rho_a^{-1}\ \nabla p \quad (5)$$

for the pressure driving force. With the last three relations Equation (2) then assumes the modified 'atmospheric' structure

$$\vec{I}_p = - \rho D_p \nabla x_p + \rho x_p B_p \{- m(\frac{\rho_p}{\rho_a} \vec{F}_e + \vec{F}_f) + \vec{F}_c + \vec{F}_v + \vec{F}_t\} \quad (6)$$

which indicates that a real flux-force relation for the aerosol current density $\vec{I}_p$ has been obtained where $\vec{I}_p$ consists of a sum of purely thermodynamic driving forces which do not explicitly depend on $\vec{I}_p$. In particular, it is notable that the flux equation for $\vec{I}_p$ now explicitly includes Coulomb force effects which follow as a consequence of the elimination of the dependent inertia forces. Note further the absence of gravity and buoyancy forces in (6) which corresponds to the'reversible character'which one expects for the activity of mechanical external forces. Detailed studies on this problem (e.g. Herbert, 1980) show that only in cases of mechanical equilibrium, gravity and buoyancy forces can play a significant role in the irreversible particle flux. However, the condition of mechanical equilibrium is not adopted here.

For the remainder of this study the flux relation (6) is employed in the approximate form

$$\vec{I}_p = - \rho\ D_p \nabla x_p + \rho\ x_p B_p\ (\vec{F}_c + \vec{F}_v + \vec{F}_t) \quad (7)$$

i.e. without the transport contributions by an external electric field and friction in the air which are neglected in comparison to the phoretic and Coulomb force effects. Assuming constant density as well as constant diffusivity and mobility coefficients, Equation (7) is applied to the dynamical constraints

$$\nabla \cdot \vec{I}_p = 0 \quad \text{and} \quad \nabla \cdot (\vec{F}_c + \vec{F}_v + \vec{F}_t) = 0 \quad (8)$$

in order to derive from it a characteristic steady state distribution of the particles concentration x_p around a collector drop of radius a. Solving (8) as spherically symmetric problem, i.e. with respect to the distance coordinate R, and considering the boundary conditions

$$x_p(R = \infty) = x_o \ , \qquad x_p(R = a) = 0 \ , \tag{9}$$

one obtains after a series of elementary, straightforward calculations the following $x_p(R)$ distribution:

$$x_p(R) = x_o \ \frac{\exp\left[\frac{B_pC}{D_pf_pa}\left(1 - \frac{a}{R}\right)\right] - 1}{\exp\left(\frac{B_pC}{D_pf_pa}\right) - 1} \ . \tag{10}$$

In particular, this x_p-distribution explicitly depends on the diffusity D_p and mobility B_p of the particles as well as on the factor C predicted as

$$C = R^2 \ (\vec{F}_c + f_v\vec{F}_v + f_t\vec{F}_t) \cdot \vec{e}_R \quad (\vec{e}_R = \text{unit vector}). \tag{11}$$

(11) exhibits that C is explicitly determined as function of the Coulomb and phoretic driving forces $\vec{F}_c$, $\vec{F}_v$ and $\vec{F}_t$.

Note in addition the slightly modified structure of (10) and (11) by a set of mean ventilation coefficients f_p, f_v, f_t introduced to compensate missing forces due to the fact that in natural atmospheric conditions collector drops move and the temperature, vapor and particle distributions around moving drops are therefore not spherically symmetric. Ventilation coefficients have been suggested and used in Wang et al. (1978) as a quasi-first order approximation for missing hydrodynamic effects within the idealized concept of equation (8). Thereby, f_p is used as ventilation factor for Brownian diffusion of particles while f_v and f_t are used as ventilation factors for the vapor and temperature fields. In numerical evaluations of the present study ventilation effects remain disregarded setting $f_v = f_t = f_p = 1$, for convenience.

With Equations (7), (9) and (10) the collection mechanism is directly coupled with the integral (total) particle flux

$$J_p = \left| \int_{\Omega(a)} \vec{I}_p \cdot d\vec{\Omega} \right| = \frac{4\pi B_pC\rho x_o}{\exp\left(\frac{B_pC}{D_pf_pa}\right) - 1} \tag{12}$$

through the surface area $\Omega(a) = 4\pi a^2$ of the collector drop. Employing (12) to determine the collection kernel K, it is a reasonable approach for the problem at hand to write

$$K \approx \frac{J_p}{\rho x_o} = \frac{4\pi B_pC}{\exp\left(\frac{B_pC}{D_pf_pa}\right) - 1} \tag{13}$$

which indicates in comparison to the general definition of diffusive kernels (Pruppacher and Klett, 1978) that K is determined by various effects of the particle transport but without any influence by the collector drop motion.

The absence of drop properties within K is hypothesized as realistic in so far as the removal process of aerosol particles by drop collection is only considered for particles with radii $r \ll a$ (cf. also Figure 1-3).

With the expedient K-relation (13) a number of special cases and practical problems associated with the particle scavenging process can be treated. In view of numerical evaluations it is of particular importance to specialize the characteristic diffusion constants D_p, B_p and C in (13). For the diffusivity and mobility of particles of radius r moving in air of viscosity η_a it is useful to employ the relation

$$D_p = B_p kT = \frac{(1 + \alpha(r)\ N_{kn}(r))\ kT}{6\pi\eta_a r} \tag{14}$$

representing the formulas of Einstein and Stokes improved by the slip-flow correction factor $1 + \alpha N_{kn}$, where k = Boltzmann constant, $\alpha = 1.26 + 0.40 \exp(-1.10\ N_{kn}^{-1})$ and $N_{kn} = \lambda_a/r$ is the Knudsen-number with λ_a being the mean free path length of air molecules. Numerical values of η_a and λ_a used in this study are given in the list below. Equation (14) may be evaluated for the two following main cases:

1. Particles move due to Brownian motion, only, i.e. Coulomb and phoretic forces are omitted, C = 0. Then (13) readily turns over to the simple Brownian collection kernel

$$K = 4\pi a\ f_p D_p \tag{15}$$

represented by the curves in Figure 1.

2. Phoretic and Coulomb forces are not omitted. In consistency with Grover et al. (1977) and Wang et al. (1978) where this case was argued in detail, C can be expressed with fair accuracy by the complex relationship

$$C = Q_r Q_a - \frac{7.15\ \pi\eta_a\ D_v f_v\ a\ (\rho_v - \rho_{v,a})\ r}{1 + \alpha\ N_{kn}}$$

$$- \frac{2.40\ \pi\eta_a\ (K_a + 2.5K_p\ N_{kn})\ K_a f_t a\ (T - T_a)\ r}{(1 + 3N_{kn})(K_p + 2K_a + 5K_p\ N_{kn})\ p} \tag{16}$$

The curves in Figure 2 and 3 show collection kernels evaluated according to (13) with C from (16) and the following coefficients:

drop and particle charges: $Q_a = \pm\ 2.0\ a^2$, $Q_r = \mp\ 2.0\ r^2$
in e.s. units and a and r in 10^{-2} m.

$k = 1.38\times10^{-23}$ mkg s^{-2} K^{-1}

$\lambda_a = 7.24\times10^{-8}$ m

$K_p = 4.19\times10^{-1}$ Jm^{-1} s^{-1} $°C^{-1}$ thermal conductivity of aerosol particles

$K_a = 2.44\times10^{-2}$ - " - - " - - " - of air

$\eta_a = 1.77\times10^{-5}$ kg m^{-1} s^{-1} air viscosity

$D_v = 0.26\times10^{-4}$ m^2 s^{-1} vapor diffusivity

$f_v = f_t = f_p = 1$

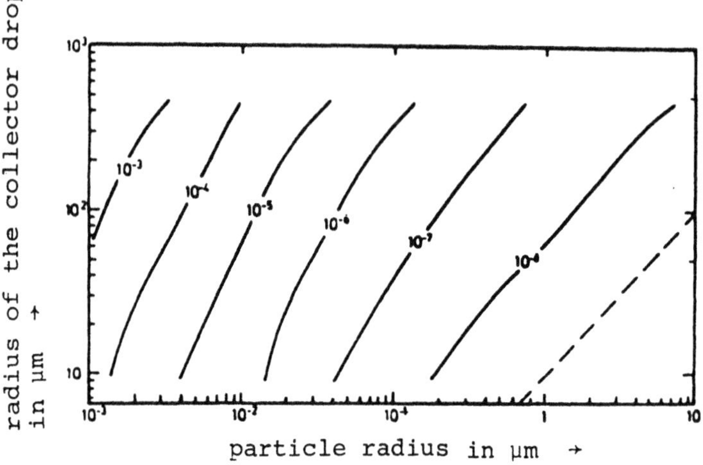

Fig. 1: Brownian collection kernel (cm^3s^{-1}) at ambient temperature T = 10° C and pressure p = 900 mb

Relative humidity and temperature at the drop surface are expressly determined in C by assuming that a drop with radius a is in thermodynamic equilibrium with its ambient air at temperature T_a. Thus, $\rho_{v,a}$ and T_a follow from a seperate calculation scheme where a drop-psychrometer equation for the temperature and a Köhler-Kelvin type equation for the saturation pressure at the drop surface must be solved. The explicit treatment of these equations is not further discussed in this paper.

Note also that all numerical calculations of this study were uniquely performed for drop radii 10 ≤ a ≤ 500 µm and particle radii 0.001 ≤ r ≤ 10 µm as well as for an ambient temperature T=10° C, an ambient pressure p=900 mb and an ambient humidity RH = 50% and 101%.

A comparison of the curves in Figure 1 to 3 exhibits the effects caused by phoretic and Coulomb forces relative to the effect caused by Brownian diffusion, only. We note the following important features of the computed collection kernels:

1) At the same atmospheric conditions T=10° C, p=900 mb and RH=50% one finds in all cases for a given particle radius r that the kernel increases with increasing drop radius a where the Brownian kernel throughout provides the smallest values at any arbitrary particle size. This result was expected and can easily be understood by looking at (13) and (15) and considering that K represents the effective volume of collection per unit time.

2) For a given drop radius a and a variable particle radius r the K-curves show a quite different behavior in the various cases. While the inclusion of phoretic forces leads in supersaturated air (RH=101%) to a strong decrease, it leads in subsaturated air (RH=50%) to a considerable increase of the kernel compared with the Brownian kernel. Note particularly in subsaturated air with RH≐50% the increase of the minimum values about the factor 10^3-10^4 to $K=10^{-5}$ cm^3s^{-1} in comparison to $K=10^{-8}$-10^{-9} cm^3s^{-1} in the Brownian case.

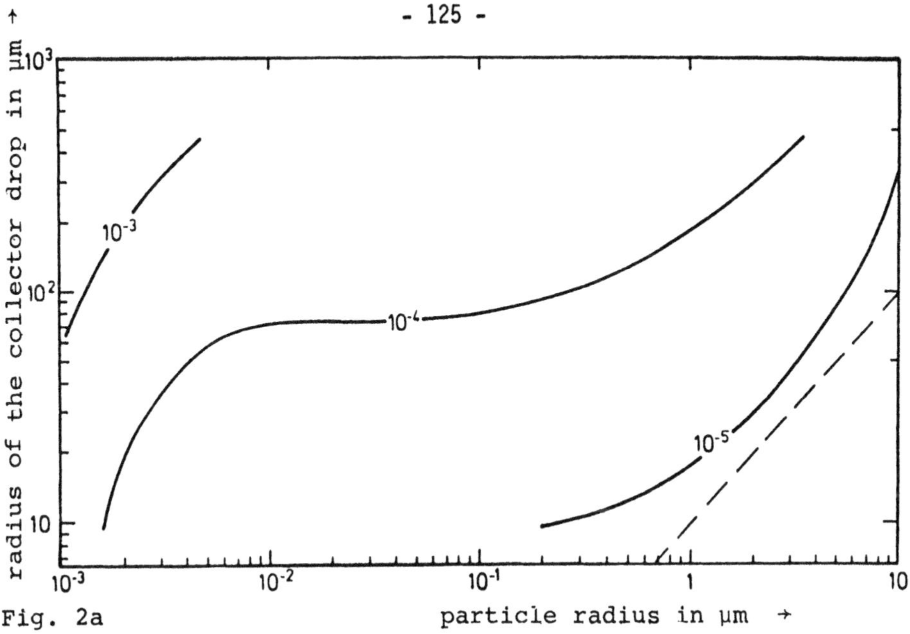

Fig. 2a

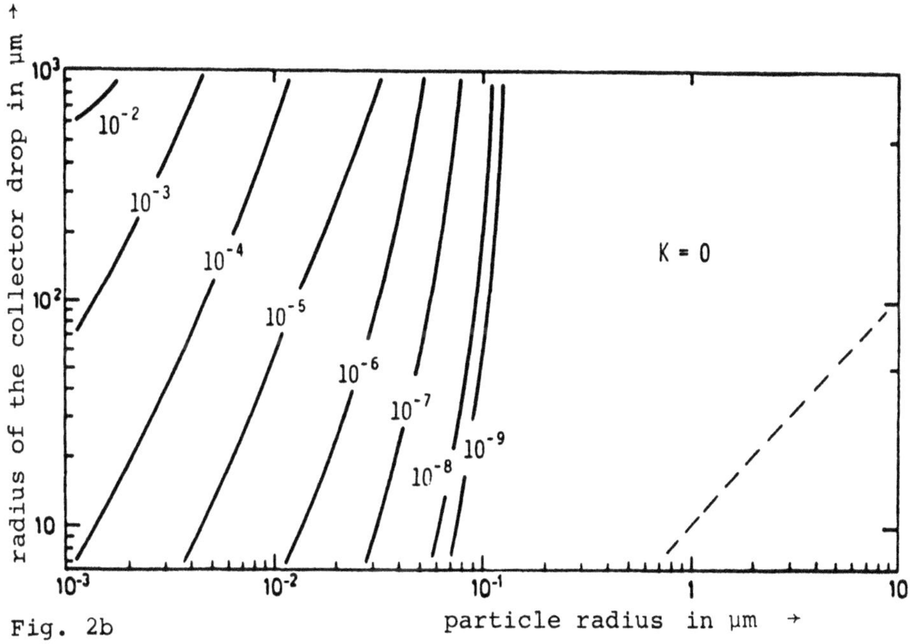

Fig. 2b

Collection kernel ($cm^3 s^{-1}$)
for water drops and particles of various sizes

Driving forces : Brownian diffusion, phoretic forces

Ambient air conditions: temperature 10° C, pressure 900 mb

Relative humidity : 50% (at Fig.2a) and 101% (at Fig.2b)

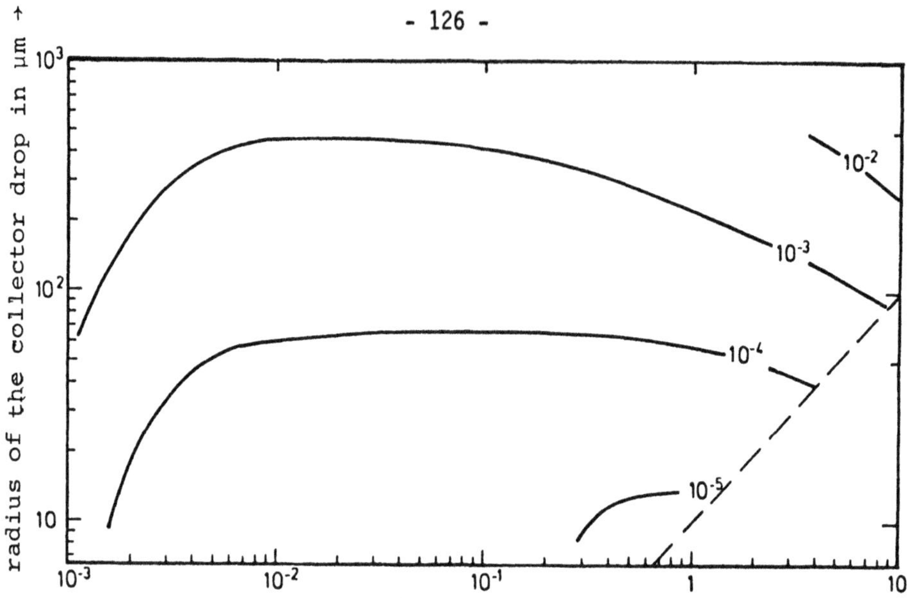

Fig. 3a particle radius in μm →

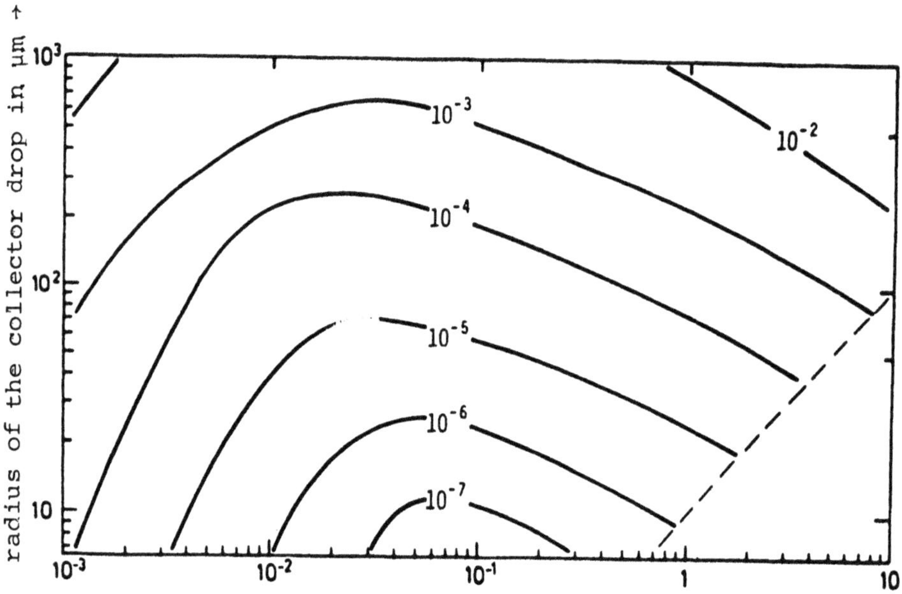

Fig. 3b particle radius in μm →

Collection kernel ($cm^3 s^{-1}$)
for water drops and particles of various sizes

Driving forces: Brownian diffusion, phoretic forces, Coulomb forces (electric charge on drop and particles: $Q(r) = \pm 2.0$ e.s.u. $cm^{-2}\, r^2$, $Q(a) = \mp 2.0$ e.s.u. $cm^{-2}\, a^2$)

Ambient air conditions: temperature 10° C, pressure 900 mb

Relative humidity : 50% (at Fig.3a) and 101% (at Fig.3b)

Conversely, in supersaturated air with RH = 101%, phoretic forces cause an extremely steep slope of the kernel to vanishing small values beginning at particle radii about 10^{-2} - 10^{-1} µm. It is not easy to find a specific reasoning of this astonishing and apparently contradictory phoretic behavior since the quantitative impact on the collection strongly varies with relative humidity and other ambient parameters. In principle, however, the fact is decisive that the combined force by diffusio- and thermophoretic effects and the force by Brownian diffusion may have the same direction (as in the case of RH=50%) or opposite direction (as in the case of RH=101%).

3) The influence on the particle collection caused by the attraction effect of Coulomb forces is shown in Figure 3. From these results follows that Coulomb force attraction affects a considerable strengthening of the collection process for particles with radii $r \geq 10^{-2}$ µm. In particular it follows that all curves from about $r \approx 10^{-2}$ µm show a distinct slope up to particle radii of about $r \approx 10$ µm. This feature is of a principle importance and indicates, with respect to a given drop radius, the particular behavior that the kernel gradually increases with increasing particle size. One may conclude that this impact in the collection of greater particles corresponds to the 'inertial character' which one expects for the diffusion and collection activity of Coulomb forces as a consequence of the substitution of inertia forces by combining (2) with (4).

3. CONCLUSIONS

It is suggested to treat the collection mechanism of aerosol particles by water drops in air as a problem of the particle diffusion flux. Explicit theoretical and numerical calculations seem to provide realistic collection kernels for practical applications.

General diffusion laws and specialized hydrodynamic equations formulate a model which describes Brownian diffusion, diffusio- and thermophoretic as well as Coulomb force effects and which is particularly applicable to predict aerosol scavenging kernels with respect to spherically symmetric conditions. The model indicates that, in comparison to Brownian and phoretic driving forces being real gradient forces, the physical behavior of Coulomb driving forces in the collection mechanism is similar to that of inertia forces. This treatment is plausible and reveals discrepancies and misconceptions in earlier formulations with respect to the description of inertia and Coulomb forces. Also, it must be realized that the inclusion of Coulomb forces according to Herbert (1978) is not justified.

Acknowledgement

Many thanks are due to K. Beheng and Miss M. Roos for their collaboration in the problems of this theme and for providing the figures of this paper.

4. REFERENCES

BEARD, K,V., and S.N. GROVER, 1974: Numerical collision efficiencies for small raindrops, J. Atmos. Sci., 31, 543-550

GROVER, S.N., 1976 a: A numerical determination of the effect of electric fields and charges on the efficiency with which cloud drops and small raindrops collide with aerosol particles, Pure Appl. Geophys., 114, 509-520

----, 1976 b: The electrostatic force between a charged conducting sphere and a charged dielectric sphere in an arbitrary oriented electric field, Pure Appl. Geophys., 114, 521-539

----, and K.V. BEARD, 1975: A numerical determination of the efficiencies with which electrically charged cloud drops and small raindrops collide with electrically charged spherical particles of various densities, J. Atmos. Sci., 32, 2156-2165

----, H.R. PRUPPACHER and A.E. HAMIELEC, 1977: A numerical determination of the efficiency with which spherical aerosol particles collide with spherical water drops due to inertial impaction, and phoretic and electric forces, J. Atmos. Sci., 34, 1655-1663

HALL, D.W., and K.V. BEARD, 1975: Electrostatic force between a conducting sphere and a dielectric sphere, Pure Appl. Geophys., 113, 515-524

HERBERT, F., 1978: A theoretical model to describe the motion of aerosol particles due to the combined action of inertia, Brownian diffusion and phoretic and electric forces, J. Atmos. Sci., 35, 1744-1750

----, 1980: Prigogine's diffusion theorem and its application to atmospheric transfer processes, Part 1: The governing theoretical concept , Contr. Atmos. Phys., 53, 181-203

WANG, P.K., S.N. GROVER and H.R. PRUPPACHER, 1978: On the effect of electric charges on the scavenging aerosol particles by cloud and small raindrops, J. Atmos. Sci., 35, 1735-1743.

Elemental Particle Size Distributions of the Tropospheric Aerosol, Determined by PIXE

Petra Metternich, H. W. Georgii, K. O. Groeneveld

Johann Wolfgang Goethe-Universität, Frankfurt am Main 90, FRG

A b s t r a c t

Particle Induced X-Ray Emission (PIXE) has been used for the analysis of the elemental particle size distributions of the tropospheric aerosol and its elemental composition as a function of altitude. Data obtained for the first time in these studies will be discussed.

1. INTRODUCTION

Particle Induced X-Ray Emission (PIXE) has been developed into a powerful analytical tool. First introduced by Johansson and Johansson (1976), the increasing number of practicle applications demonstrates its usefulness and its capability of being competitive with the standard methods. Therefore, especially for meteorology and aerosol physics, PIXE is an important method to yield information of environmental significance (Metternich et al., 1981). PIXE is now capable of rapid routine determination of the elemental concentrations in aerosol size fractions as collected by single orifice impactors over short periods of time. This enables the detailed study of complex relationships between detected elements. The variation of particulate abundance with particle size may give clues to processes at sources and during particle transport and removal.

2. BASIC PRINCIPLES OF PIXE

The main features of PIXE can best be described by comments to fig. 1. The atoms considered are bombarded by

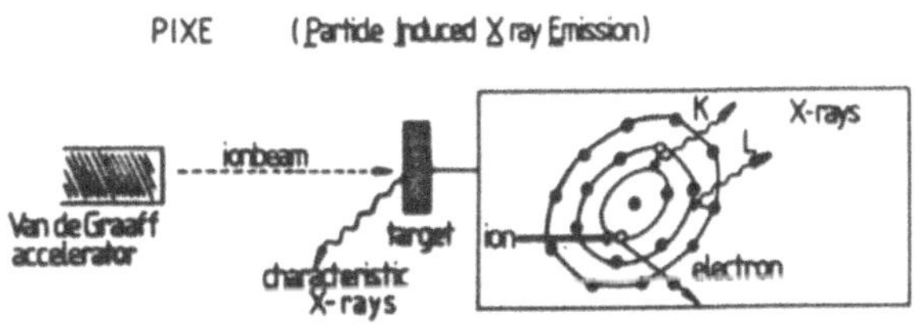

fig. 1 Schematic diagram of the basic principles of PIXE

protons, produced in an accelerator with energies between 1-10 MeV per nucleon. The incident particles will interact with the atoms ejecting electrons and thereby ionizing this atom. The holes thus created in the inner electron shells, e.g. in a K- or L-shell, are subsequently filled by one after the outer electrons and the transition energy can be transfered to characteristic K or L X-radiation. The characteristic X-rays detected are proportional to Z^2 (Z=atomic number) according to Moseleys Law, so the qualitative analysis is done; quantitative data are obtained by evaluation of line intensities (Folkmann et al., 1974).

The advantages of the PIXE are the following:
- PIXE is multielemental. Up to 20 elements can be determined simultaneously.
- PIXE is non destructive.
- The sensitivity is high, 10^{-6}-10^{-12}g minimum detectable amount.
- Very small samples can be analyzed with full sensitivity. Using the microprobe technique, even microstructures can be analyzed.
- PIXE is fast. The proton exposure time period for most samples is in the order of 5-15 min; if the resulting spectra are transfered to a computer they can be analyzed and a plot be available a few minutes after a run.
- No chemical break down of the sample is necessary.

3. EXPERIMENTAL

Relatively little information about the elemental composition of aerosols in the troposphere is available, information which is necessary for the comprehension of processes at sources and during particle transport and removal (Gillette and Blifford, 1971). Most of the reported measurements have been made near the ground. To extend our present knowledge of the aerosol trace element particle size distributions in the troposphere, we have collected samples at ground and in altitudes up to 6 km in polluted and remote areas.

3.1 SAMPLING PROCEDURE

The collecting device was a single orifice impactor of the PIXE design (PIXE International Corporation, Tallahassee, Florida). Aerodynamic particle cutoff diameter for the stages are: 7, > 16; 6, > 8; 5, > 4; 4, > 2; 3, > 1; 2, > 0.5; 1, > 0.25 µm diameter and behind back-up filter < 0.25 µm diameter at 1l/min flowrate. For aircraft instrumentation an intake nozzle-designed to preserve isokinetic conditions - is mounted below the fuselage, while the impactor and the pumping system were placed inside the cabin of the DO28 Aircraft. Aerosol collections at ground level were made at Frankfurt/Main,Germany and St. Moritz, Switzerland to obtain typical elemental particle size distributions of a polluted and a remote area. Aircraft measurements were performed in the Bavarian highlands at three altitudes from 1.3 to 5.5 km.

3.2 ANALYSIS

The PIXE analysis of the aerosol samples were carried out at the Institut für Kernphysik, Johann Wolfgang Goethe-Universität, Frankfurt/Main, Germany, using the 2.5 MV Van de Graaff accelerator. The irradiation facilities are standard and have been described by Metternich et al. (1980).

4. RESULTS

The data obtained by ground measurements are summarized in fig. 2. For 15 elements - Si, S, Cl, K, Ca, Ti, V, Cr, Mn, Fe, Ni, Cu, Zn, Br, Pb - the particle size distributions

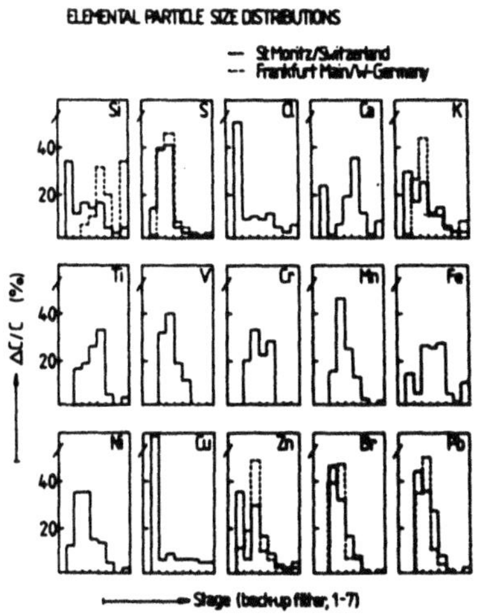

fig. 2 Elemental particle size distributions determined in St. Moritz and Frankfurt/Main
C = elemental concentration

could be determined, presented for the remote area St. Moritz in fig. 2. The variation of particulate abundance with particle size gives clues to the processes at sources. Soil dust, e.g. Si, Ca, Fe, and sea spray droplets are found preferentially at the large end of the size spectrum, which is typical of particles formed at ordinary temperatures by dispersion of solids or liquids. Some organic matter, particulate sulphur compounds, lead and bromine from automobile exhaust, are found mainly at the small end of the size spectrum, characteristic of particles formed by condensation or gas-to-particle-conversion. These elements show a significant difference in their particle size distributions at the two sampling locations. Coagulation processes of small particles in polluted areas with high particle concentrations as shown by Georgii and Kaller (1970) turn out

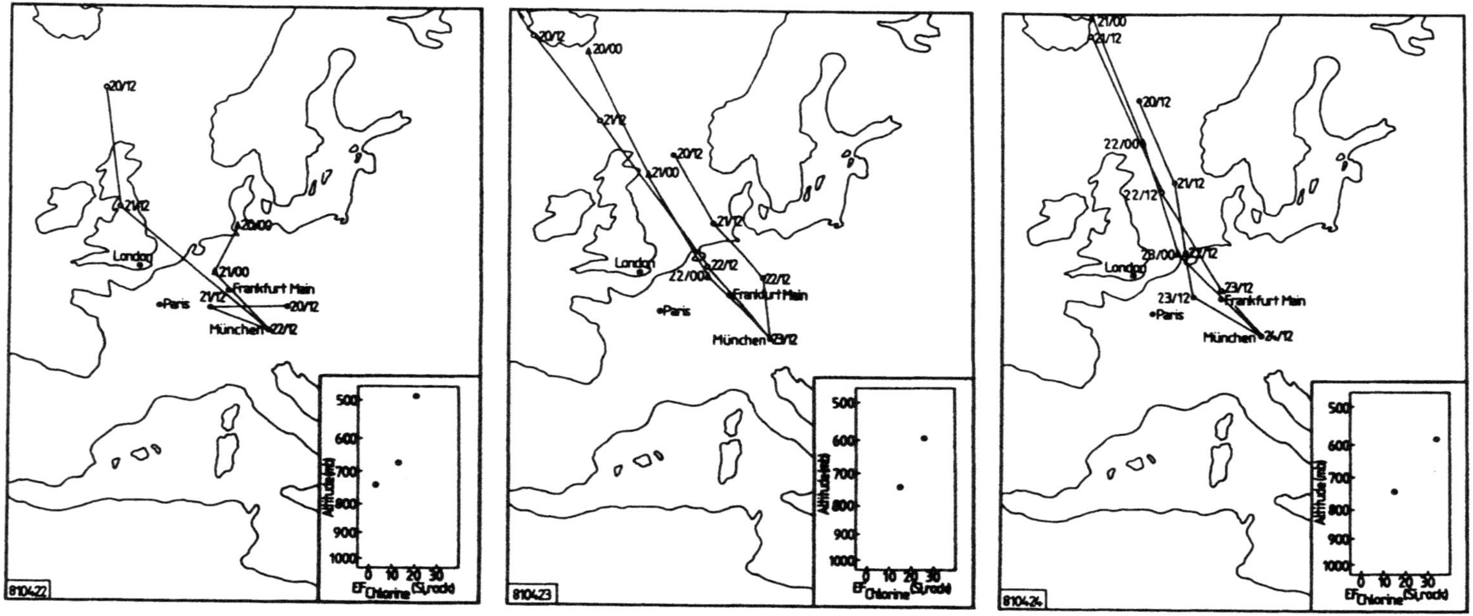

fig. 3a-c Air mass trajectories and Chlorine enrichment factors, calculated for the research flight on April 22 to 24, 1981

● 1000 mb, ○ 850 mb, ▲ 700 mb trajectories

datecode: 22/12 = April 22, 1981 12 GMT

by a displacement to the large end of the size spectrum, also given in fig. 2.

Investigations of transport processes of aerosol particles in the atmosphere are based on the knowledge of the particle size distributions of particular elements at various sampling sites. A first research flight on three days in April 81 yields unusual high chlorine concentrations in all flightlevels with large particle size orientated distributions similar to those found at marine sampling sites suggesting a sea-spray chlorine contribution. In addition, enrichment factors for each flightlevel have been calculated, using Silicon as crustal reference element and average crustal rock as reference material as described by Rahn (1976). Inspection of these data reveals an increase in chlorine contribution from the lowest altitude up to the higher levels according to air mass history as indicated by calculated air mass trajectories (fig.3a-c) (Schütz, 1981). The highest chlorine contributions are obtained in air masses which have passed over the Northatlantic and North Sea as observed in the upper flightlevels where wind speeds of more than 40 kn involve a transport of sea-spray particles from the ocean to the sampling site; while the enrichment of the air with chlorine, however, is somewhat lower when the air masses have passed continental areas.

5. CONCLUSIONS

Particle Induced X-Ray Emission has been used for the study of the elemental composition of the tropospheric aerosol. Typical particle size distributions for several elements could be determined in ground measurements at different sampling locations, giving clues to processes and sources. The composition of tropospheric aerosol particles as a function of altitude and transportmechanism in the atmosphere have been studied in aircraft measurements. High chlorine concentrations observed with size distributions similar found at marine sampling sites indicate a sea-spray contribution at the sampling site as supported by airmass trajectory analysis and calculation of enrichment factors. The multielemental character of the PIXE method enables in further studies a detailed analysis of transport processes for all the other elements not mentioned here.

Acknowledgement

The authors thank V. Johannes and L. Schütz (Institut für Meteorologie der Johannes Gutenberg Universität Mainz, Germany) for providing the air mass trajectories.

6. REFERENCES

FOLKMANN, F. et al., 1974: Proton Induced X-Ray Emission as a Tool for Trace Element Analysis; Nucl. Instr. and Methods 116, 487-499

GEORGII, H.-W. and KALLER, R.S., 1970:Über die Inaktivierung von Gefrierkernen durch Koagulation mit Aitkenkernen; Berichte des Instituts für Meteorologie und Geophysik der Universität Frankfurt/Main Nr. 21

GILETTE, D.A. and BLIFFORD, I.H., Jr., 1971: Composition of Tropospheric Aerosols as a Function of Altitude; Journal of the Atm. Sci. 28, 1199-1210

JOHANSSON, S.A.E. and JOHANSSON, T.B., 1976: Analytical Application of Particle Induced X-Ray Emission; Nucl. Instr. and Methods 137, 473-516

METTERNICH, Petra et al., 1980: Elementanalyse atmosphärischer Aerosole in Frankfurt/Main für den Zeitraum 1966-1977 mittels PIXE; Berichte des Instituts für Meteorologie und Geophysik der Universität Frankfurt/Main Nr. 42

METTERNICH, Petra et al., 1981: Time Dependence of the Elemental Composition of Atmospheric Aerosols over a Time Period of 12 Years; Nucl. Instr. and Methods 181, 431-434

RAHN, K.A., 1976: The Chemical Composition of the Atmospheric Aerosol; Technical Report Graduate School of Oceannography, University of Rhode Island Kingston

SCHÜTZ, L., 1981: Isobar trajectories program (unpublished)

Investigation on the Saturation-Spectra of Ice-Nuclei

D. Stein and H. W. Georgii

Institute of Meteorology and Geophysics, J. W. Goethe-University Frankfurt/Main, FRG

A b s t r a c t

The activity of the atmospheric ice-nuclei is not only dependent on the activation temperature, but also on the environmental saturation ratios over water and ice, respectively. The relation between the ice-nuclei concentration and the saturation at different temperatures was investigated with a low pressure diffusion chamber. With this instrument it is possible to measure the saturation spectra at different temperatures on the same filter. The comparison of the saturation spectra received from parallel filters showed a good reproducibility. Aerosol-samples were taken in the Swiss Alps and in the environment of Frankfurt/M. and analysed. A constant ratio was found between the slopes of the ice-supersaturation spectra and the respective slopes of the temperature spectra.

1. INTRODUCTION

Previous measurements of ice-nuclei concentrations performed at a temperature of -18°C and water saturation showed a range of concentration between o.o1 ice-nuclei/l and 1o ice-nuclei/l (Georgii and Stein, 1981).

Measurements of ice-nuclei saturation spectra, measured by Gagin (1972), by Huffman (1973) and during the III. Int. Workshop on Ice-Nucleus Measurements 1975, (Vali 1976), yielded a saturation dependency of the ice-nuclei concentration at a constant activation temperature, which had a maximum range of three orders of magnitude.

This large range shows the importance to investigate not only the temperature spectra of ice-nuclei but also the saturation spectra.

The aim of our present work was the measurement of the activation of atmospheric ice-nuclei dependance of supersaturation and temperature.

A necessary condition for these measurements is a very accurate control of activation temperature and vapor pressure. The filter method offers good presuppositions to carry out such measurements, because the filters sampled at different sampling sites can be developed under controlled and comparable conditions in a laboratory. The disadvantage of this method, the well known volumeeffect was reduced so far that a detection in the range of the sampled volumes was not possible when a low pressure diffusion chamber was used. This type of chamber first designed at our institute and presented on the III. Int. Workshop on Ice-Nucleus Measurements, 1975 (Meyer and Gravenhorst 1976) shows no volumeeffect, because the diffusion velocity is enhanced in a low pressure system.

With the experience obtained by using this chamber a new low pressure diffusion chamber was built to reach the necessary reproducibility for these measurements by improving the closeness of the low pressure system, the temperature control and measuring system and the preparation of the filters. (A discription of this new chamber is prepared for publication.) This new chamber permits the measurement of saturation spectra of ice-nuclei at different temperatures of activation with one exposed filter. A repetition of these measurements with the same filter shows no change of the activation behaviour of the ice-nuclei.

2. EXPERIMENTAL RESULTS

An example of the results is shown in Fig.1.

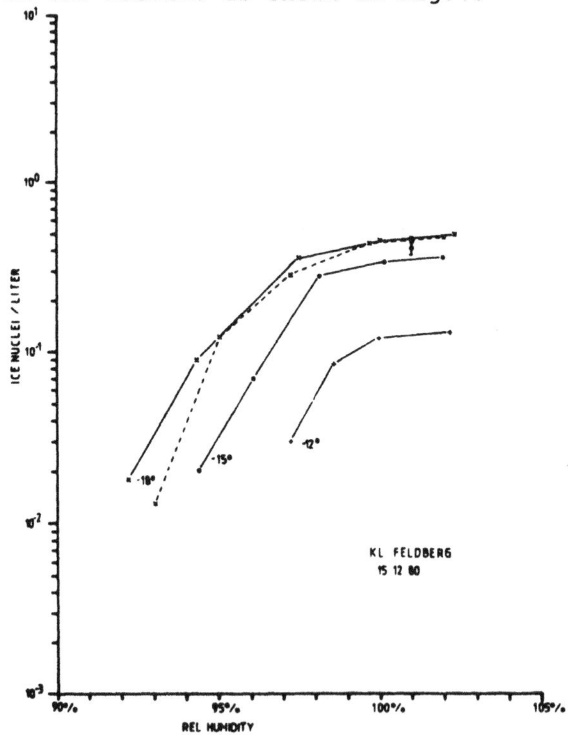

Fig.1 A comparison of a set of 4 parallel sampled filters

In this case, the ice-nuclei of the first and second filter of a set of four parallel exposed filters were activated five times at -18°C and 1% water-supersaturation marked by two points. The variation of these measurements is shown by a vertical bar. The ice-nuclei of the third filter, marked with dashed lines, were activated at -18°C at five different increasing humidities. The ice-nuclei of the fourth filter, here marked with full lines, were activated also at five different increasing humidities, but also at three different temperatures, at -12°C, -15°C and -18°C.

To avoid a memory effect, the filtersurface was heated up to +15°C with a remaining relative humidity $< 1\%$ for some minutes between the measurements of the single saturation spectra.

A comparison of the ice-nuclei concentration at -18°C and 1% water-supersaturation shows, that there is no change in the activation behaviour when the activation is repeated at different humidities and temperatures.

Usually, two parallel exposed filters were developed to control the variations of such a set and to recognize possible errors of single measurements.

Fig.2 shows the saturation spectra at three different temperatures of two simultaneously exposed filters, plotted with dashed lines. The full lines between correspond to the computed means of the single spectra of both filters. These means are used in the further treatment of our results.

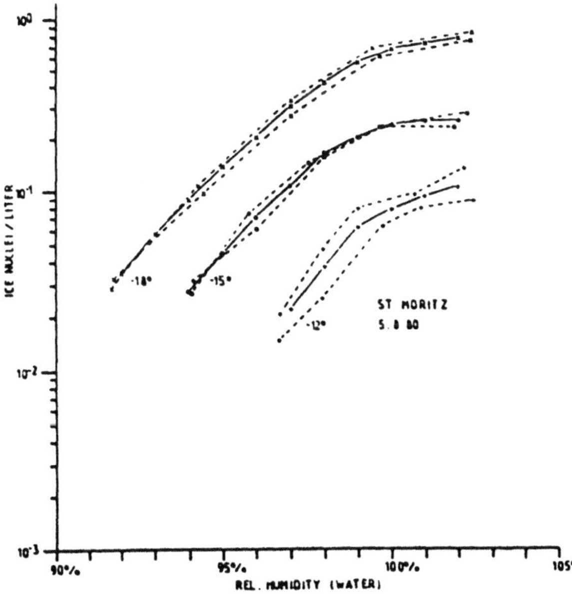

Fig.2 Saturation spectra of two parallel sampled filters at -12°C, -15°C and -18°C (---) and the mean spectra of both filters (——)

Plotting the mean spectra of Fig.2 in relation to ice-supersaturation the three spectra taken at -12°C, -15°C and -18°C lie on one line. Only the values measured at water-supersaturation deviate from this line. The deviation can clearly be seen in a double logarithmic diagram (Fig.3). In this figure the water-saturation for each temperature is marked by two vertical bars.

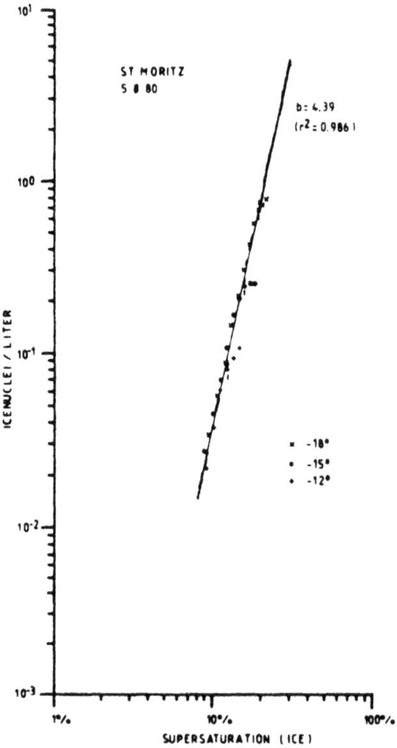

Fig.3 Ice-supersaturation spectrum at -12°C, -15°C and -18°C with the computed slope b

In many cases there was no further increase of ice-nuclei concentration after exceeding water saturation, in other cases there was a smaller increase when water saturation was exceeded. Measurements made up to now did not show a clear relationship to a specific parameter of the constitution of the aerosols, the environmental conditions or the sampled volume.

In the evaluation of our measurements we did not regard results gained at supersaturations above water-saturation.

As shown by Huffman (1973) it is possible to describe the increase of the ice-nuclei concentration at increasing ice-supersaturation by the function $N = C \cdot S_i^b$, where S_i is the supersaturation over ice, C is a constant, b the slope factor and N the number of ice-nuclei per liter. To compute the slope b, the three spectra measured at different temperatures were summarized to one ice-supersaturation spectrum. The straight line in Fig.3 corresponds to the computed ice-supersaturation spectra of the three plotted single spectra. The good correlation shown in this example is generally existent, the mean r^2 of all ice-supersaturation spectra measured up to now yield o.93. The results of the slopes measured at different sampling sites varied between 3.1 and 12.5.

Samples from St. Moritz, Switzerland, had a mean slope of 4,3; samples from Mt.Kleiner Feldberg near Frankfurt, Germany, had a mean slope of 6.o and in Frankfurt the mean slope was 6.8.

These results show an increase of the slope from unpolluted to polluted regions.

First filters exposed airborne in an altitude of 4ooo ft above Southern Germany showed a mean slope of 6.4 which was also found on a mountain-station in 3ooo ft. The trajectories showed that the airmasses during the flight period moved slowly within two days over North-Western Europe and simultaneously sulfate concentrations measured confirmed the polluted condition of these airmasses. In higher altitudes up to 18ooo ft the mean slope was 4.9.

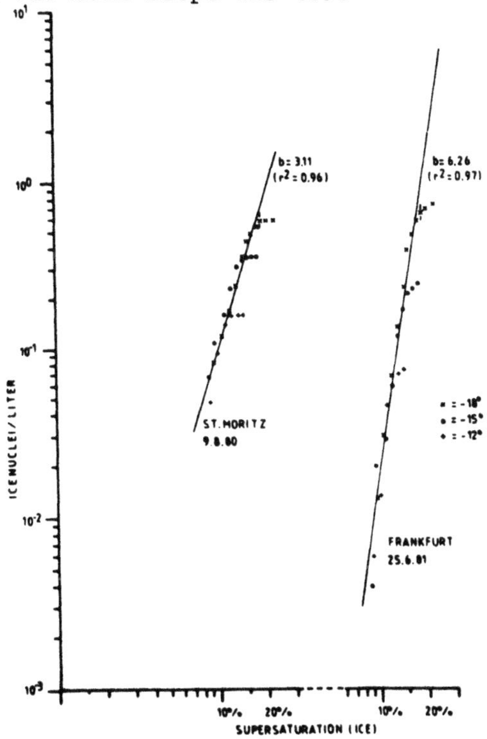

Fig.4 A typical ice-supersaturation spectrum of St.Moritz (Switzerland) and of Frankfurt/Main (Germany).

Fig.4 shows an ice-supersaturation spectrum of a sample, collected at St.Moritz with a flat slope of 3.11 and a spectrum of Frankfurt with a steep slope of 6.26 for comparison. While there is only a small difference in concentration at high ice-supersaturations, the difference increases towards low supersaturations.

The mean concentration at -18°C and water saturation measured in Frankfurt was o.6 ice-nuclei/l and in St. Moritz, Switzerland,o.55 ice-nuclei/l.The mean concentration at -12°C and water saturation in Frankfurt was o.o4 ice-nuclei/l and in St. Moritz o.1 ice-nuclei/l. Taking -18°C as reference, the concentration was too high in St. Moritz or too low in Frankfurt at high temperatures respectively at low ice-supersaturations.

Parallel measurements of the aerosol spectra yielded that neither the concentration nor the size distribution are of major influence on the ice-nuclei saturation spectra. The chemical composition and surface of the aerosol seems to be decisive.

3. THE RELATION BETWEEN SUPERSATURATION AND TEMPERATURE SPECTRA

Taking as demonstrated in Fig.5 for the extreme slopes 2 and 8 of an ice-supersaturation spectrum, the concentrations at water saturation and at different temperatures, a temperature spectrum may be computed.

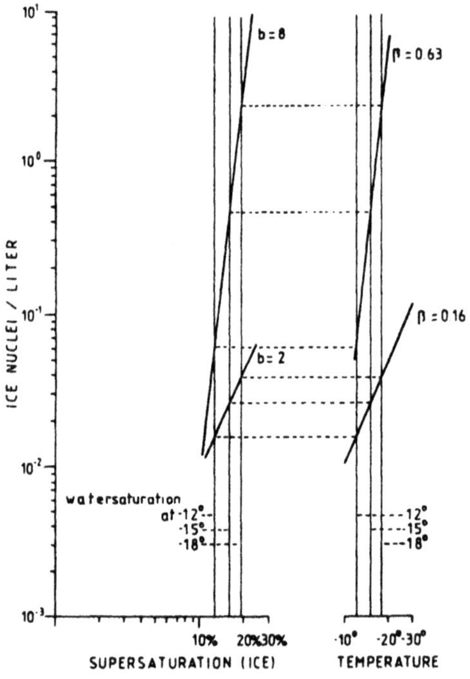

Fig.5 Representation of the relation between two ice-supersaturation spectra and two temperature spectra

For comparison with former published values, the exponential function $N = N_o \cdot \exp(\beta(T_o - T))$, where β is the slope factor, N_o a numerial factor, (T_o-T) the degree of supercooling, and N the number of ice-nuclei per liter, is fitted to the increase of ice-nuclei concentration with decreasing temperature, according to Fletcher (1962). For each slope b of the ice-supersaturation spectra a respective slope β of the temperature spectra is received. In the investigated range between -1o°C and -2o°C the computed ratio b/β was 12.78 for all slopes b from 1 to 1o by taking the ice-nuclei concentrations at water saturation for each temperature. Taking the ice-nuclei concentration at 98% rel. humidity the ratio was 1o.92. The ratio of both slopes of the ice-nuclei concentration at 98% is plotted as a dashed line in Fig.6.

The symbols in Fig.6 correspond to the measured slopes b of the ice-supersaturation spectra and the respective measured slopes β of the temperature spectra of different sampling sites.

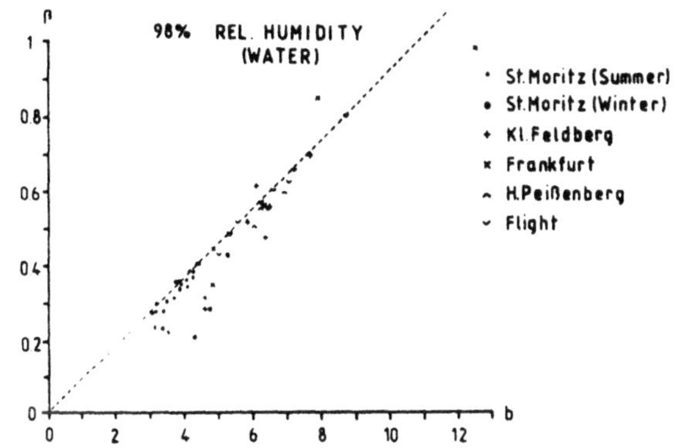

Fig.6 Computed ratio slope b to β at 98% rel. humidity(---) and the measured slopes b and β

4. CONCLUSIONS

4.1 The results of the investigations show that the low-pressure diffusion chamber permits to measure saturation-spectra of ice-nuclei with the same filter at different temperatures. The results are reproducible.

4.2 The relation between humidity and activity of ice-nuclei leads to an increase of the ice-nuclei concentration according to a power-law. This increase is independant of the temperature as long as water-saturation is not exceeded.

4.3 The slope of the ice-supersaturation spectra becomes smaller at water supersaturation, independant of the actual concentration of nuclei on the filter. This is in confirmation of the results gained by Tanaka (198o).

4.4 The change of the slope at water saturation shows that the computed water-saturation is in agreement with the actual water-saturation on the filter-surface.

4.5 The concentration of activated ice-nuclei on the filter at different water subsaturations but at equal ice-supersaturations coincide. Therefore the rel. humidity below 1oo% water-saturation does not play a dominant role in the activation process of ice-nuclei.

4.6 The inclination of the slope increases from unpolluted to polluted areas in agreement with Huffman (1973). The inclination of the slope decreases with increasing altitude in the free atmosphere.

4.7 There is a close correlation between the measured and computed ratio of ice-supersaturation- and temperature-spectra only in unpolluted air.

4.8 At low supersaturation the concentration of activated ice-nuclei is relatively low in polluted air. Following a hypotheses by Georgii (1963) and Georgii and Kaller (197o) it is assumed that surface-poisoning leads to a partial deactivation of ice-nuclei by trace-gas adsorption or coagulation with inactive Aitkin-nuclei. This mechanism may also influence the ice-supersaturation spectrum.

5. REFERENCES

Fletcher, N.H., 1962: Physics of rain clouds. Cambridge University Press London

Gagin, A., 1972: The effect of supersaturation on the ice crystal production by natural aerosols. J. Rech.Atmos.6, 175-185

Georgii, H.W., 1963: Investigation on the deactivation of inorganic and organic freezing nuclei. Z. Angew. Math. Physik 14, 5o3-51o

Georgii, H.W. und Kaller, R.S., 197o: Über die Inaktivierung von Gefrierkernen durch die Koagulation mit Aitkenkernen. Berichte des Instituts für Meteorologie und Geophysik, Universität Frankfurt/Main Nr.36

Georgii, H.W. und Stein, D., 1981: Die Konzentrationsverteilung atmosphärischer Gefrierkerne in der unteren und mittleren Troposphäre. Met. Rdsch. 34, 137-143

Huffman, P., 1973: Supersaturation dependence of ice nucleation by deposition for silver iodide and natural aerosols. Res. Rept. No. AR 1o8. Sept. 1973, Dept. of Atmospheric Resources, University of Wyoming, Laramie, Wyoming

Meyer, D. and Gravenhorst, G., 1976: A low pressure diffusion chamber. The Third International Workshop on Ice Nucleus Measurements. In: G.Vali, Proceedings, University of Wyoming, Laramie, Wyoming, 128-148

Tanaka, T., 198o: Ice nucleating activity and the mode of action of volcanic ash ejected from Mt. Usu in Hokkaido. An improved method to remove hygroscopic materials collected on a membrane filter. Papers in Meteorology and Geophysics, Vol.31, Nos. 3/4 pp. 153-171

Vali, G., 1976: The Third International Workshop on Ice-Nucleus Measurements. Proceedings, University of Wyoming, Laramie, Wyoming

The Determination of Absorptive Properties of Atmospheric Aerosol Particles Using Photoacoustic Spectroscopy

Bernd Wrobel

Institut für Meteorologie der Universität Mainz, FRG

A b s t r a c t

Chopped light produces an acoustic signal in a closed cell when it is absorbed by a sample of material within this cell. This effect is used to develop a method for measuring the absorption coefficient of aerosol particles. The possibilities and limits of the method are discussed.

1. INTRODUCTION

Many meteorological problems demand knowledge of the absorption coefficient of aerosol particles disperged in the atmosphere as shown by Fischer (1970) and Andre et al (1981). It is required for instance to calculate the radiant flux through the atmosphere. Furthermore the molecular composition of these aerosol particles is unknown till now. The photoacoustic spectroscopy shall be used to determine the absorption coefficient and via the absorption coefficient the molecular composition of aerosol particles.

In 1880 A.G. Bell discovered that a periodically interrupted light beam incident on a solid, liquid or also on a gaseous material produces an audible sound. This effect is the base of the photoacoustic spectroscopy, particularly suitable to investigate opaque substances.

2. DESCRIPTION OF THE PHOTOACOUSTIC EFFECT

The sample material is placed in a closed cell containing a nonabsorbing gas, for instance air at atmospheric pressure and room temperature. Chopped monochromatic light penetrates a window of the cell and incidents on the sample. The light is partly absorbed by the sample and partly reflected. The radiation absorbed by the sample is transformed in heat by nonradiative de-excitation. This produces a thermal wave in the sample which, reaching the sample surface heats a thin layer of gas, adjacent to the surface of the sample. This causes this boundary layer of air to expand, and so it acts as a cyclic piston, creating a pressure variation within the cell. The pressure fluctuations are proportional to the heat of the sample, caused by absorptive and thermal properties of the sample at the surface. The pressure variations are detected by a microphone placed in the photoacoustic cell.

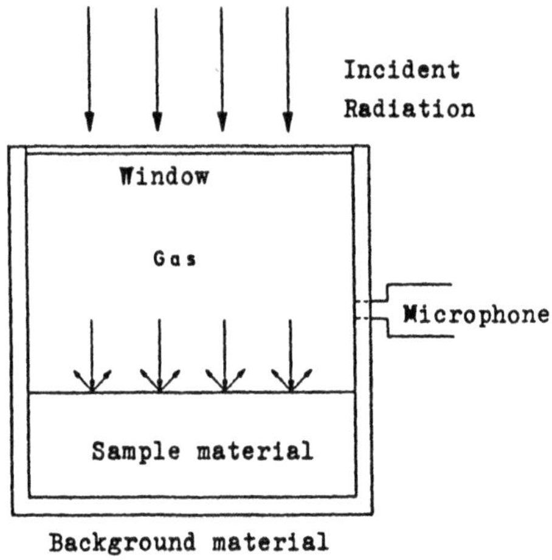

Fig. 1: The photoacoustic cell

Stray light can be neglected in photoacoustic spectroscopy. The reason for this is the high thermal conductivity and diffusivity of the walls. A theory of the process in the cell is given by Rosencwaig and Gersho (1976), but till now not yet verified.

3. CONSTRUCTION OF THE PHOTOACOUSTIC SPECTROMETER

The light source of the photoacoustic spectrometer used is a 300 Watt xenon lamp. The light passes through a mechanical chopper and then enters a monochromator. The monochromator and a filter wheel ensure quasi monochromatic radiation from 250 to 2400 nm. A beam splitter

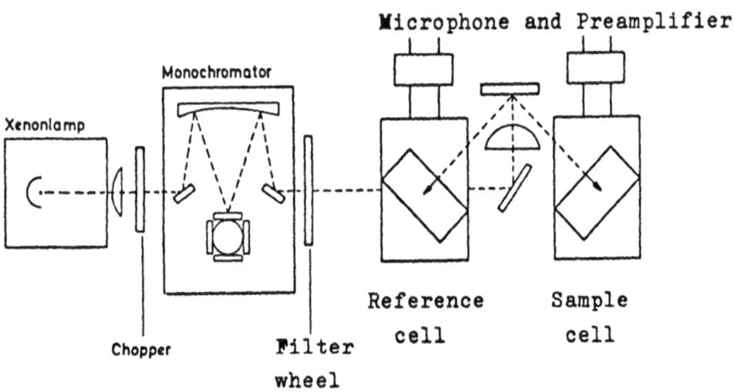

Fig. 2: The construction of the photoacoustic spectrometer

divides the light beam into two beams of equal intensity. Each beam is then directed into separate symetric cells: a reference cell and a sample cell. The reference material is carbon black.

The ratio of both signals is a signal largely independent of properties of the spectrometer itself and of the spectrum of the xenon lamp. It depends on the thermal properties of reference and sample and the absorption coefficient of the sample. It does not depend on the absorption coefficient of the reference material because of the signal saturation of the carbon black.

4. THE EFFECT OF SIGNAL SATURATION

In the case of signal saturation the local heating by absorption of radiation takes place immediately at the surface of the sample. The incident radiation is completely absorbed and the signal does not depend any more on the absorption coefficient. So the increase of the signal is limited. Prehn (1979) has shown that saturation is reached when the absorption coefficient of a sample is about 10^4 cm^{-1}. All parts of a spectrum lying within the region of saturation will be cut off then, as shown in figure 3.

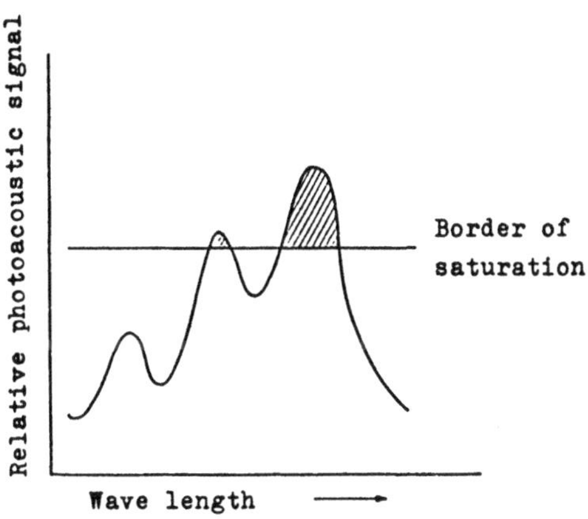

Fig. 3: The spectrum of an absorbing material being within saturation partly

It is possible to examine samples with respect to signal saturation. To do that it is necessary to vary the chopper frequency at a fixed wavelength of the light. The signal plotted versus the chopper frequency in a graph with logarithmic axes shows a straight line with negative slope. Carbon black is in the region of saturation at every wavelength of the photoacoustic spectrometer, its slope is -0.7.

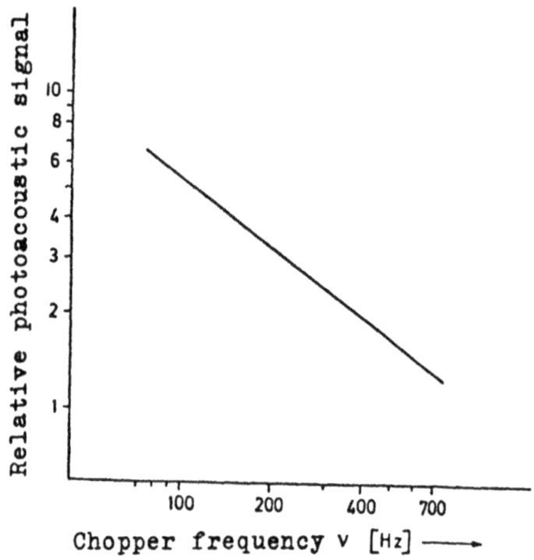

Fig. 4 : Examination of soot at 496 nm

5. EXAMINATION OF AEROSOL PARTICLES

Samples of aerosol particles were collected on the roof of the Institute of Meteorology in Mainz using an impactor with two stages. The collected particles were divided into two ranges of size, one extending from 0,1 to 1 µm radius and one with radii larger than 1 µm. The sample holders were constantly moved back and forth in front of the jet orifice of the impactor stages. So a uniform layer of particles was produced on the sample holder.

The quantity of particles necessary for investigation could only be reached by collecting several days. Till now it was not possible to examine the samples during short term weather conditions. Nevertheless an influence of the season on the sample could be shown. Samples collected in winter during inversions always are within saturation and have an absorption coefficient larger than 10^4 cm^{-1}. The samples made in springtime and summer are not saturated.

All samples of aerosol particles examined up to now don't have a selective absorption. It follows that they are grey absorbers. An explanation for that is the high concentration of soot in the aerosol particles of the anthropogenic polluted atmosphere in Mainz. It is obvious that a potential absorption of other components of the aerosol particles is suppressed by the portion of soot. So substances with known absorption spectra were diluted by soot to find out at which concentration of soot a spectrum could be detected. It was not possible to discover a limiting mixing ratio up to now. Some mixtures were diluted down at 10^5 parts of the mixture itself to one part of soot. Also the spectral behaviour of mixtures without soot have been investigated.

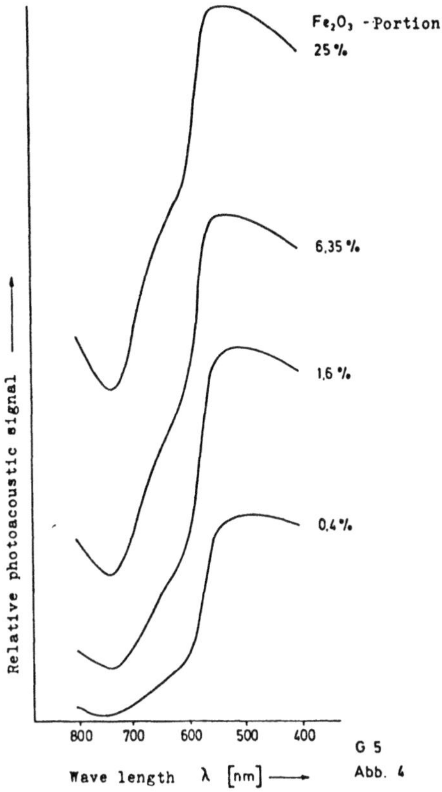

Fig. 5: Spectra of mixtures of Fe_2O_3 and NaCl with varying composition

If such mixtures are diluted by nonabsorbing materials it is possible to estimate the concentration of the components, if their spectra are known.

6. A METHOD FOR MEASURING ABSORPTION COEFFICIENTS

As there is no reliable theory for the photoacoustic spectroscopy and because of the difficulties to determine their thermal properties of the sample material, a special method was developed for measuring the absorption coefficient of samples.

The basic requirement for the application of this method are identical experimental conditions (for example constant radiant flux densities, chopper frequency, volume and pressure of gas in the cell etc.). Given this the photoacoustic signal only depends on the rate of heating of the surface, even if there are samples of different material and structure. In this case it is unimportant how the change of temperature is achieved at the surface. The reason can be chopped light or periodical heating of the sample by an electrical resistor. In order to apply this method sample holders were used, covered by a resistor foil.

According to the conditions above radiation incident on sample holders, one covered by the sample (S) only, one covered by a resistance foil (R) and one by a resistance foil and a thin layer of

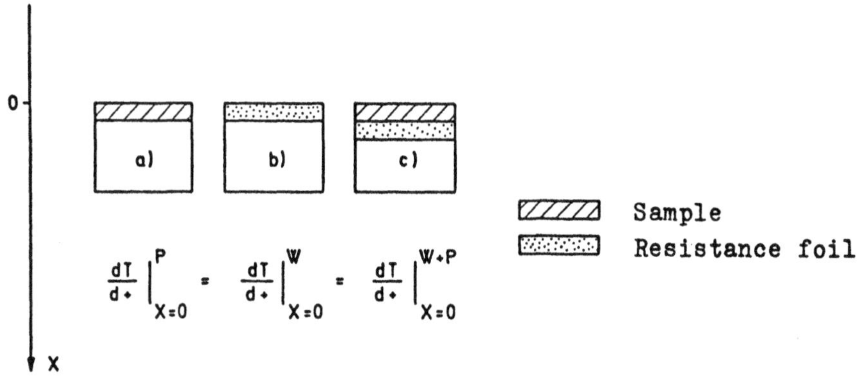

Fig. 6: Generation of the photoacoustic signal
a) by chopped monochromatic light
b) by an electrical current in a resistor foil
c) like b) with an additional layer of sample material

the sample material (R+S) (see fig. 6). The layers shall be homogeneous and boundary effects shall be negligible. A one dimensional coordinate system is used where X is the axis, vertical to the surface of the sample. X is zero where the light hits the surface of the sample. In the case of identical photoacoustic signals the following equation is valid

$$\left.\frac{d\,T}{d\,t}\right|_{X\,=\,0}^{S} = \left.\frac{d\,T}{d\,t}\right|_{X\,=\,0}^{R} = \left.\frac{d\,T}{d\,t}\right|_{X\,=\,0}^{R\,+\,S} \qquad (1)$$

starting with the equation

$$\frac{d\,Q}{d\,t} = c \cdot M\ \frac{d\,T}{d\,t} \qquad (2)$$

Q = Quantitiy of heat (cal), t = Time (sec)
M = Mass of S, R or R+S (g), T = Temperature (deg Kelvin)
c = Specific heat (cal g^{-1} deg^{-1})

and the electrical equivalent of heat

$$Q = \text{Factor}\ R \cdot I^2 \cdot t \qquad (3)$$

R = Resistance (Ohm) I = Intensity of current (Amp)

follows

$$\frac{dT}{dt} = \frac{1}{C} R \cdot I^2 \quad (4)$$

C = Heat capacity (cal deg^{-1})

This equation is valid corresponding to the surface of R and R+S

$$\left.\frac{dT}{dt}\right|_{X=0}^{R} = \frac{1}{C_R} R_R \cdot I_R^2 \quad (5)$$

$$\left.\frac{dT}{dt}\right|_{X=0}^{R+S} = \frac{1}{C_R + C_S} R_{R+S} \cdot I_{R+S}^2 \quad (6)$$

Further on

$$\frac{1}{\alpha} \left.\frac{\partial T}{\partial t}\right|^{S} = - \frac{1}{\varkappa} \frac{\partial F_\lambda(0)}{\partial X} \quad (7)$$

α = Thermal diffusivity (cm^2 sec^{-1})
$\varkappa$ = Thermal conductivity (cal cm^{-1} sec^{-1} deg^{-1})
F_λ (0) = Radiant flux density at X = 0 (W cm^{-2})

F_λ (0) is given by the Lambert Beer law. According to (1) and using equations (5), (6), (7), the equation to determine the absorption coefficient β is

$$\beta = \frac{(R \cdot I^2)_{R+S} \, (\chi - 1)}{\chi \cdot A \cdot d \cdot F_\lambda(0)} \quad (8)$$

$$\chi = \frac{I_{R+S}^2 \cdot R_{R+S}}{I_R^2 \cdot R_R} \quad (9)$$

β = Absorption coefficient (cm^{-1})
A = Area of the sample which must be heated (cm^2)
d = Thickness of the sample (cm)

All quantities on the right side of the equation (8) are measurable.

7. THE MEASURING TECHNIQUE

The currents can be determined during the experiment, and there is no difficulty to find out A, R_R, R_{R+S}. The radiant flux density of the radiation incident on the sample is measured by photodiodes with an integral procedure. The photodiodes are calibrated comparing with an Eppley thermopile and a monochromator. The thickness of the sample can be determined by a geometrical measuring technique using a raster scan microscope.

The experimental process is as follows. At first the photoacoustic signal of the sample on the resistance foil is produced at a fixed wavelength, then obscuring the sample, the same signal by electrical heating. Finally the same photoacoustic signal is created by electrical heating of the resistance foil, not covered by particles.

8. CONCLUSIONS

The procedure has been tested with ZnSe, a substance of known absorption coefficient, $B = 1{,}08\ cm^{-1}$ at 514,5 nm as shown by Hordvik and Schlossberg (1977). The scattering of the results of these tests has been too large. The reasons were incalculable contact variations during the electrical heating. To avoid this a new designed photoacoustic cell is in construction.

An analysis of the molecular chemical composition of aerosol particles is difficult because of their portion of carbon black. It would be necessary to develop a method to separate the soot from the bulk material. Then it is possible to determine the other components of aerosol particles by means of their photoacoustic spectra.

9. REFERENCES

ANDRE, K., DLUGI, R., SCHNATZ, G., 1981: Absorption of visible radiation by atmospheric aerosol particles, fog and cloud water residues. J. of the Atm. Sciences, Jan. 81, Vol. 38, No. 1, 141-155.

FISCHER, K., 1970: Bestimmung der Absorption von sichtbarer Strahlung durch Aerosolpartikel. Beitr. Phys. d. Atm., 43. Bd., 244-254.

HORDVIK, A., SCHLOSSBERG, H., 1977: Photoacoustic technique for determining optical absorption coefficients in solids. Appl. Opt., Jan. 1977, Vol. 16, No. 1, 101-107.

PREHN, H., 1979: Photoakustische Spektrospkopie: Ergänzung oder Alternative zur Optischen Spektroskopie. GIT Fachzeitschrift für das Laboratorium, 23. Jhrg., Heft 4, April 1979, 281-289.

ROSENCWAIG, A., GERSHO, A., 1976: Theory of the photoacoustic effect with solids. J. of Appl. Phys., Vol. 47, No. 1, Jan. 76, 64-69.

GPSR Compliance

The European Union's (EU) General Product Safety Regulation (GPSR) is a set of rules that requires consumer products to be safe and our obligations to ensure this.

If you have any concerns about our products, you can contact us on ProductSafety@springernature.com

In case Publisher is established outside the EU, the EU authorized representative is:

Springer Nature Customer Service Center GmbH
Europaplatz 3
69115 Heidelberg, Germany

Batch number: 10682411

Printed by Printforce, the Netherlands